electronics four

Hayden Electronics One-Seven Series

Harry Mileaf, Editor-in-Chief

electronics one — Electronic Signals □ Amplitude Modulation □ Frequency Modulation □ Phase Modulation □ Pulse Modulation □ Side-Band Modulation □ Multiplexing □ Television Signals □ Navigation Signals □ Facsimile □ Heterodyning □ Harmonics □ Waveshaping

electronics two — Electronic Building Blocks □ Basic Stages □ The Power Supply □ Transmitters □ Receivers □ UHF □ Telemetry □ Television □ Radar and Sonar □ RDF □ Radio Navigation □ Radio Control □ Quad Sound

electronics three — Electron Tubes □ Diodes □ Triodes □ Tetrodes □ Pentodes □ Multielement Tubes □ Gas-Filled Tubes □ Phototubes □ Electron-Ray Indicators □ Cathode-Ray Tubes □ UHF and Microwave Tubes □ Magnetrons □ Klystrons □ The Traveling-Wave Tube

electronics four — Semiconductors □ P-N Diodes □ Switching Diodes □ Zener Diodes □ Tunnel Diodes □ N-P-N and P-N-P Transistors □ Photodevices □ Field-Effect Transistors □ JFET's and MOSFET's □ Integrated Circuits □ Mesa and Planar Transistors

electronics five — Power Supplies □ Rectifiers □ Filters □ Voltage Multipliers □ Regulation □ Amplifier Circuits □ A-F, R-F, and I-F Amplifiers □ Video Amplifiers □ Phase Splitters □ Follower Amplifiers □ Push-Pull Amplifiers □ Limiters

electronics six — Oscillators □ Sinusoidal and Nonsinusoidal Oscillators □ Relaxation Oscillators □ Magnetron Oscillators □ Klystron Oscillators □ Crystal Oscillators □ Modulators □ Mixers and Converters □ Detectors and Demodulators □ Discriminators

electronics seven — Auxiliary Circuits □ AVC, AGC, and AFC Circuits □ Limiter and Clamping Circuits □ Separator, Counter, and Gating Circuits □ Time Delay Circuits □ Radio Transmission □ Antennas □ Radiation Patterns □ R-F Transmission Lines

electronics four

HARRY MILEAF EDITOR-IN-CHIEF

HAYDEN BOOKS
A Division of Howard W. Sams & Company
4300 West 62nd Street
Indianapolis, Indiana 46268 USA

SECOND EDITION
SEVENTH PRINTING — 1989

International Standard Book Number: 0-8104-5957-4
Library of Congress Catalog Card Number: 75-45505

Printed in the United States of America

preface

This volume is one of a series designed specifically to teach electronics. The series is logically organized to fit the learning process. Each volume covers a given area of knowledge, which in itself is complete, but also prepares the student for the ensuing volumes. Within each volume, the topics are taught in incremental steps and each topic treatment prepares the student for the next topic. Only *one* discrete topic or concept is examined on a page, and *each* page carries an illustration that graphically depicts the topic being covered. As a result of this treatment, neither the text nor the illustrations are relied on solely as a teaching medium for any given topic. Both are given for *every* topic, so that the illustrations not only complement but reinforce the text. In addition, to further aid the student in retaining what he has learned, the important points are summarized in text form on the illustration. This unique treatment allows the book to be used as a convenient review text. Color is used not for decorative purposes, but to accent important points and make the illustrations meaningful.

In keeping with good teaching practice, all technical terms are defined at their point of introduction so that the student can proceed with confidence. And, to facilitate matters for both the student and the teacher, key words for each topic are made conspicuous by the use of italics. Major points covered in prior topics are often reiterated in later topics for purposes of retention. This allows not only the smooth transition from topic to topic, but the reinforcement of prior knowledge just before the declining point of one's memory curve. At the end of each group of topics comprising a lesson, a summary of the facts is given, together with an appropriate set of review questions, so that the student himself can determine how well he is learning as he proceeds through the book.

Much of the credit for the development of this series belongs to various members of the excellent team of authors, editors, and technical consultants assembled by the publisher. Special acknowledgment of the contributions of the following individuals is most appropriate: Frank T. Egan, Jack Greenfield, and Warren W. Yates, principal contributors; Peter J. Zurita, S. William Cook, Jr., Steven Barbash, Solomon Flam, and A. Victor Schwarz, of the publisher's staff; Paul J. Barotta, Director of the Union Technical Institute; Albert J. Marcarelli, Technical Director of the Connecticut School of Electronics; Howard Bierman, Editor of *Electronic Design;* E. E. Grazda, Editorial Director of *Electronic Design;* and Irving Lopatin, Editorial Director of the Hayden Book Companies.

HARRY MILEAF
Editor-in-Chief

contents

transistors and semiconductor diodes

Transistors and semiconductor diodes perform essentially the same jobs that electron tubes do in electronic equipment. They have become very important in electronics because of their many advantages over electron tubes. Transistors are much smaller and lighter than tubes. As a result, transistorized equipment is small and weighs very little. Equipment that was heavy, bulky, and permanently mounted can now be portable and miniaturized.

Transistors do not have to be heated as do electron tubes. Thus, equipment power supplies and circuit components can be made smaller, simpler, and cheaper. Transistorized equipment operates much cooler than electron-tube equipment, so that the cooling system for a complex piece of equipment is not needed or is simple. The solid transistor is much more rugged than the relatively delicate electron tube, so that shock and vibration is less of a problem. Transistors are easier to store and last longer on the shelf. One of the biggest advantages of the transistor over the electron tube is that the transistor has a long equipment life, whereas the electron tube accounts for more than half of all electronic equipment failures.

Transistors, though, do have some disadvantages. Ordinary transistors cannot handle as much power as ordinary tubes. There are high-power transistors, but they are specially made. Also transistors are very sensitive to temperature changes and radiation, and it is more difficult to manufacture transistors with consistent characteristics. By and large, however, the transistor's advantages outweigh its disadvantages; transistors are continually being improved and used more and more in commercial, industrial, and military electronics.

Transistors are small and light

Transistors do not have to be heated and run cooler

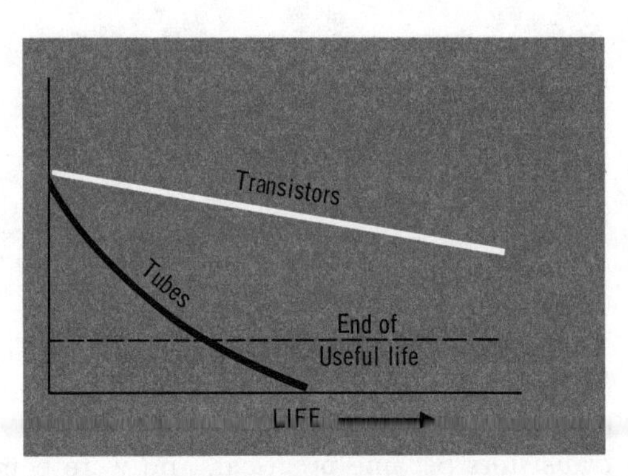

Transistors last much longer than tubes

early history

Semiconductors, and particularly semiconductor diodes, are not really new to electronics. The old crystal detector that was used in the early days of radio was a semiconductor diode; and the old copper oxide and selenium rectifiers, which are still in use today, are also semiconductor diodes. Even transistors can no longer be considered new. They were first developed in 1948 at Bell Telephone Laboratories by John Bardeen, William Shockley, and W. H. Brittain. They were looking for a *solid-state* device whose resistance could be changed in a manner similar to that of the electron tube. The original term used to describe the device they developed was *trans*fer res*istor*, which was then shortened to transistor.

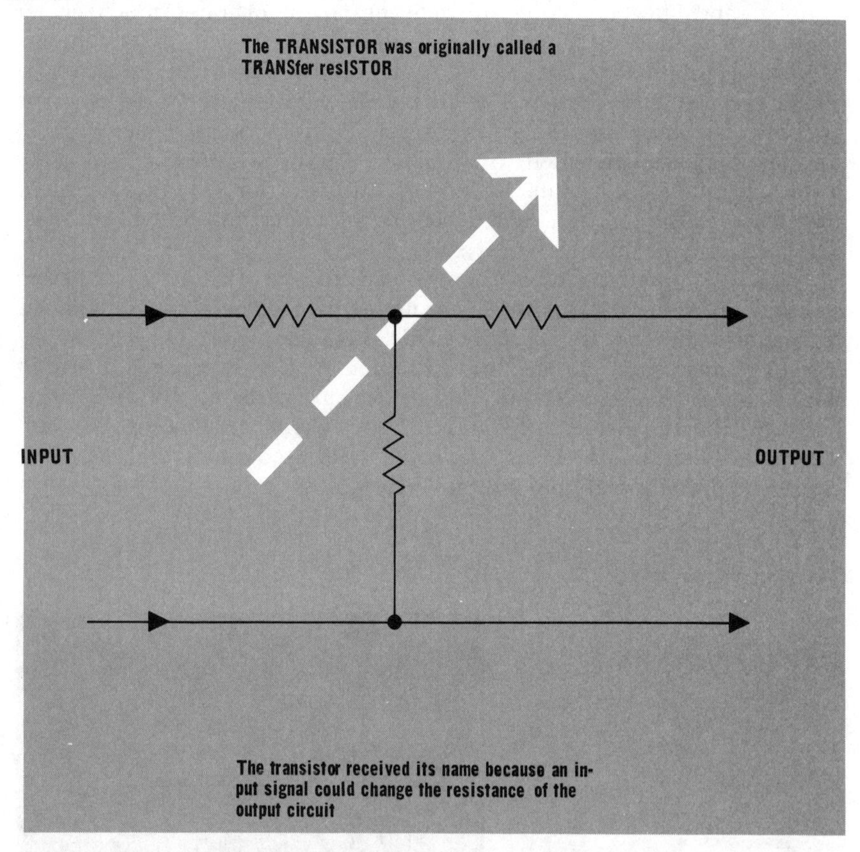

The first transistors were expensive and difficult to control. Because of this, they were at first used mostly experimentally; but as time went on, they were improved and reduced in cost. During the late 1950's, transistors became practical, and were being used in many applications that had used the electron tube. Today, transistors are continually being improved and are being put to more and more uses.

atomic and electron theories

To understand how transistors and other semiconductors work, you should have a good knowledge of atomic and electron theories. The following is a review of their principles. As you know, all matter is composed of *compounds* or *elements*. The elements are the *basic* materials found in nature. When elements are combined to form a new material, we have a compound. The smallest particle that a compound can be reduced to and still retain its properties is a *molecule*. The smallest particle that an element can be reduced to and still retain its properties is called an *atom*. When elements are combined to form compounds, the atoms of the elements join to form the compound molecules. There are thousands of compounds, but there are only slightly more than 100 elements, from which all matter is made. The elements are listed in the table with their atomic number.

THE NATURAL ELEMENTS

Atomic Number	Name	Symbol	Atomic Number	Name	Symbol	Atomic Number	Name	Symbol
1	Hydrogen	H	32	Germanium	Ge	62	Samarium	Sm
2	Helium	He	33	Arsenic	As	63	Europium	Eu
3	Lithium	Li	34	Selenium	Se	64	Gadolinium	Gd
4	Beryllium	Be	35	Bromine	Br	65	Terbium	Tb
5	Boron	B	36	Krypton	Kr	66	Dysprosium	Dy
6	Carbon	C	37	Rubidium	Rb	67	Holmium	Ho
7	Nitrogen	N	38	Strontium	Sr	68	Erbium	Er
8	Oxygen	O	39	Yttrium	Y	69	Thulium	Tm
9	Fluorine	F	40	Zirconium	Zr	70	Ytterbium	Yb
10	Neon	Ne	41	Niobium (Columbium)	Nb	71	Lutetium	Lu
11	Sodium	Na	42	Molybdenum	Mo	72	Hafnium	Hf
12	Magnesium	Mg	43	Technetium	Tc	73	Tantalum	Ta
13	Aluminum	Al	44	Ruthenium	Ru	74	Tungsten	W
14	Silicon	Si	45	Rhodium	Rh	75	Rhenium	Re
15	Phosphorus	P	46	Palladium	Pd	76	Osmium	Os
16	Sulfur	S	47	Silver	Ag	77	Iridium	Ir
17	Chlorine	Cl	48	Cadmium	Cd	78	Platinum	Pt
18	Argon	A	49	Indium	In	79	Gold	Au
19	Potassium	K	50	Tin	Sn	80	Mercury	Hg
20	Calcium	Ca	51	Antimony	Sb	81	Thallium	Tl
21	Scandium	Sc	52	Tellurium	Te	82	Lead	Pb
22	Titanium	Ti	53	Iodine	I	83	Bismuth	Bi
23	Vanadium	V	54	Xenon	Xe	84	Polonium	Po
24	Chromium	Cr	55	Cesium	Cs	85	Astatine	At
25	Manganese	Mn	56	Barium	Ba	86	Radon	Rn
26	Iron	Fe	57	Lanthanum	La	87	Francium	Fr
27	Cobalt	Co	58	Cerium	Ce	88	Radium	Ra
28	Nickel	Ni	59	Praseodymium	Pr	89	Actinium	Ac
29	Copper	Cu	60	Neodymium	Nd	90	Thorium	Th
30	Zinc	Zn	61	Promethium	Pm	91	Protactinium	Pa
31	Gallium	Ga				92	Uranium	U

THE ARTIFICIAL ELEMENTS

Atomic Number	Name	Symbol	Atomic Number	Name	Symbol	Atomic Number	Name	Symbol
93	Neptunium	Np	97	Berkelium	Bk	101	Mendelevium	Mv
94	Plutonium	Pu	98	Californium	Cf	102	Nobelium	No
95	Americium	Am	99	Einsteinium	E	103	Lawrencium	Lw
96	Curium	Cm	100	Fermium	Fm			

atomic particles

Although the atoms of the different elements have different properties, they all contain the same *subatomic* particles. There are a number of different subatomic particles, but only three of these are of interest in basic electronics: the *proton, the electron,* and the *neutron*. The only way that the atom of one element differs from the atom of another element is in the *number* of subatomic particles that it contains. You will learn more about this later.

The protons and neutrons are contained in the center, or *nucleus,* of the atom, and the electrons *orbit* around the nucleus. The proton is small but very heavy; it is difficult to dislodge from the nucleus. The electron is larger than the proton, but it is about 1840 times as light, and is easy to move.

The electrons and the protons are the particles that have the electrical properties. The electron has a *negative* electrical charge, and the proton has a *positive* electrical charge. These charges are *equal* and *opposite.* The law of electrical charges states that particles with *like* charges *repel* each other, and those with *unlike* charges *attract* each other. This is what keeps the electrons in orbit. The nucleus of the atom contains the positive protons, and so has a positive charge that attracts the negative electrons. The centrifugal force of the orbiting electrons counteracts the attraction of the nucleus to keep the electrons orbiting. Since electrons have like charges, they repel each other, and cause themselves to be spaced equidistant from one another.

Neutrons have no electrical charge; they are *neutral.* They are sometimes thought of as protons and electrons combined, but they are actually different particles. Usually, atoms have the same number of electrons and protons, and so they are electrically neutral. If an atom does have more electrons, it is called a *negative ion.* If it has more protons, it is called a *positive ion.*

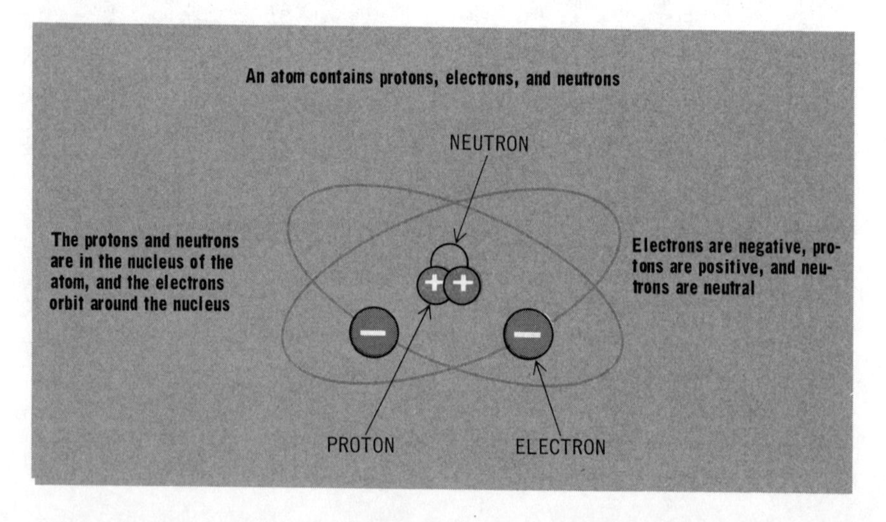

electrical charges

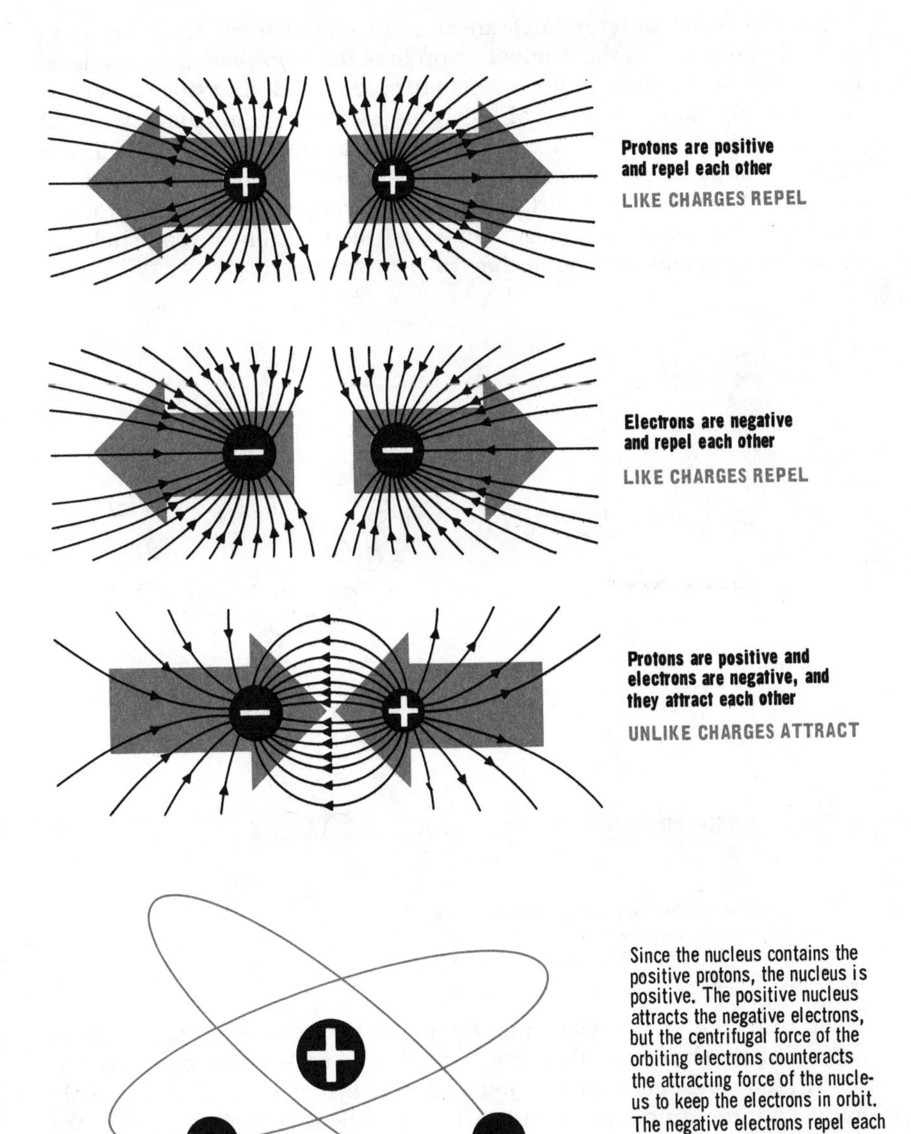

Since the nucleus contains the positive protons, the nucleus is positive. The positive nucleus attracts the negative electrons, but the centrifugal force of the orbiting electrons counteracts the attracting force of the nucleus to keep the electrons in orbit. The negative electrons repel each other, and so electrons space themselves equidistant from one another around the nucleus

orbital shells

Actually, what differentiates an atom of one element from an atom of another element is the number of protons the atom has in its nucleus. This is what the *atomic number* refers to on page 4-3. And since a neutral atom has the same number of electrons as protons, an atom with 29 protons should have 29 electrons orbiting around its nucleus. These electrons orbit in groups called *shells*. Actually, each electron has its own individual orbit, but certain orbits are grouped together to produce what is called a shell. For convenience, all the electrons in one shell are shown on diagrams as though they follow the same orbit.

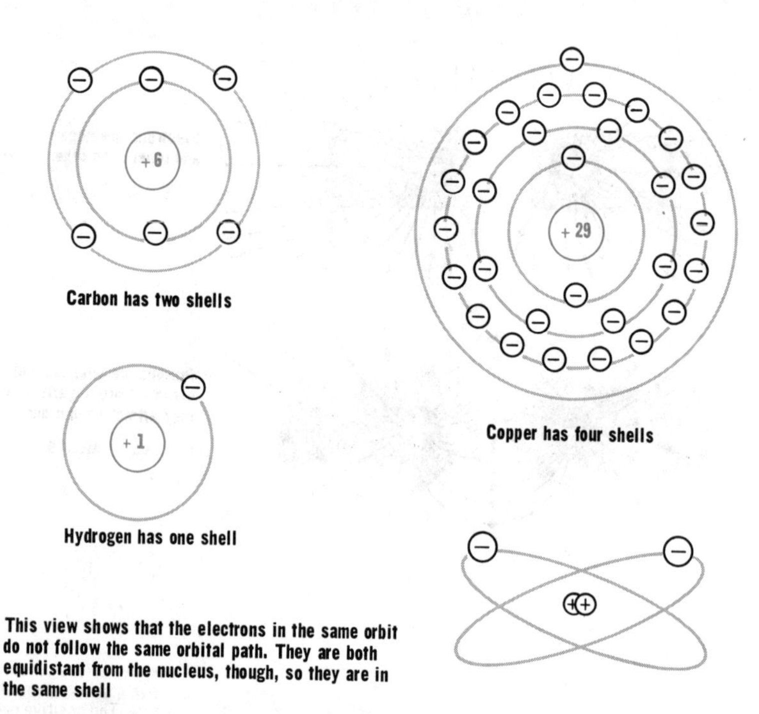

Carbon has two shells

Copper has four shells

Hydrogen has one shell

This view shows that the electrons in the same orbit do not follow the same orbital path. They are both equidistant from the nucleus, though, so they are in the same shell

For all of the known elements, there can be up to seven shells in an atom. This is shown on the table for all of the elements on page 4-7. If you study the table briefly, you will notice that each shell can only hold a certain number of electrons in orbit. The first shell, closest to the nucleus, cannot hold more than 2 electrons; the second shell cannot hold more than 8 electrons; the third no more than 18; the fourth, no more than 32; and so on. You can see that up to atomic number 10, the second shell built up to 8 electrons, and since this is the limit, a third shell had to be started for atomic number 11. For atomic numbers 11 through 18, the third shell built up to 8 electrons, and then a fourth shell started; then the third shell continued to build up to its maximum of 18 from atomic numbers 19 through 29.

the elements and their atomic shells

ELECTRON SHELLS

Atomic No.	Element	Electrons per shell 1	2	3	4	5
1	Hydrogen, H	1				
2	Helium, He	2				
3	Lithium, Li	2	1			
4	Beryllium, Be	2	2			
5	Boron, B	2	3			
6	Carbon, C	2	4			
7	Nitrogen, N	2	5			
8	Oxygen, O	2	6			
9	Fluorine, F	2	7			
10	Neon, Ne	2	8			
11	Sodium, Na	2	8	1		
12	Magnesium, Mg	2	8	2		
13	Aluminum, Al	2	8	3		
14	Silicon, Si	2	8	4		
15	Phosphorus, P	2	8	5		
16	Sulfur, S	2	8	6		
17	Chlorine, Cl	2	8	7		
18	Argon, A	2	8	8		
19	Potassium, K	2	8	8	1	
20	Calcium, Ca	2	8	8	2	
21	Scandium, Sc	2	8	9	2	
22	Titanium, Ti	2	8	10	2	
23	Vanadium, V	2	8	11	2	
24	Chromium, Cr	2	8	13	1	
25	Manganese, Mn	2	8	13	2	
26	Iron, Fe	2	8	14	2	
27	Cobalt, Co	2	8	15	2	
28	Nickel, Ni	2	8	16	2	
29	Copper, Cu	2	8	18	1	
30	Zinc, Zn	2	8	18	2	
31	Gallium, Ga	2	8	18	3	
32	Germanium, Ge	2	8	18	4	
33	Arsenic, As	2	8	18	5	
34	Selenium, Se	2	8	18	6	
35	Bromine, Br	2	8	18	7	
36	Krypton, Kr	2	8	18	8	
37	Rubidium, Rb	2	8	18	8	1
38	Strontium, Sr	2	8	18	8	2
39	Yttrium, Y	2	8	18	9	2
40	Zirconium, Zr	2	8	18	10	2
41	Niobium, Nb	2	8	18	12	1
42	Molybdenum, Mo	2	8	18	13	1
43	Technetium, Te	2	8	18	14	1
44	Ruthenium, Ru	2	8	18	15	1
45	Rhodium, Rh	2	8	18	16	1
46	Palladium, Pd	2	8	18	18	0
47	Silver, Ag	2	8	18	18	1
48	Cadmium, Cd	2	8	18	18	2
49	Indium, In	2	8	18	18	3
50	Tin, Sn	2	8	18	18	4
51	Antimony, Sb	2	8	18	18	5
52	Tellurium, Te	2	8	18	18	6

Atomic No.	Element	Electrons per shell 1	2	3	4	5	6	7
53	Iodine, I	2	8	18	18	7		
54	Xenon, Xe	2	8	18	18	8		
55	Cesium, Cs	2	8	18	18	8	1	
56	Barium, Ba	2	8	18	18	8	2	
57	Lanthanum, La	2	8	18	18	9	2	
58	Cerium, Ce	2	8	18	19	9	2	
59	Praseodymium, Pr	2	8	19	20	9	2	
60	Neodymium, Nd	2	8	19	21	9	2	
61	Promethium, Pm	2	8	18	22	9	2	
62	Samarium, Sm	2	8	18	23	9	2	
63	Europium, Eu	2	8	18	24	9	2	
64	Gadolinium, Gd	2	8	18	25	9	2	
65	Terbium, Tb	2	8	18	26	9	2	
66	Dysprosium, Dy	2	8	18	27	9	2	
67	Holmium, Ho	2	8	18	28	9	2	
68	Erbium, Er	2	8	18	29	9	2	
69	Thulium, Tm	2	8	18	30	9	2	
70	Ytterbium, Yb	2	8	18	31	9	2	
71	Lutetium, Lu	2	8	18	32	9	2	
72	Hafnium, Hf	2	8	18	32	10	2	
73	Tantalum, Ta	2	8	18	32	11	2	
74	Tungsten, W	2	8	18	32	12	2	
75	Rhenium, Re	2	8	18	32	13	2	
76	Osmium, Os	2	8	18	32	14	2	
77	Iridium, Ir	2	8	18	32	15	2	
78	Platinum, Pt	2	8	18	32	16	2	
79	Gold, Au	2	8	18	32	18	1	
80	Mercury, Hg	2	8	18	32	18	2	
81	Thallium, Tl	2	8	18	32	18	3	
82	Lead, Pb	2	8	18	32	18	4	
83	Bismuth, Bi	2	8	18	32	18	5	
84	Polonium, Po	2	8	18	32	18	6	
85	Astatine, At	2	8	18	32	18	7	
86	Radon, Rn	2	8	18	32	18	8	
87	Francium, Fr	2	8	18	32	18	8	1
88	Radium, Ra	2	8	18	32	18	8	2
89	Actinium, Ac	2	8	18	32	18	9	2
90	Thorium, Th	2	8	18	32	19	9	2
91	Protactinium, Pa	2	8	18	32	20	9	2
92	Uranium, U	2	8	18	32	21	9	2
93	Neptunium, Np	2	8	18	32	22	9	2
94	Plutonium, Pu	2	8	18	32	23	9	2
95	Americium, Am	2	8	18	32	24	9	2
96	Curium, Cm	2	8	18	32	25	9	2
97	Berkelium, Bk	2	8	18	32	26	9	2
98	Californium, Cf	2	8	18	32	27	9	2
99	Einsteinium, E	2	8	18	32	28	9	2
100	Fermium, Fm	2	8	18	32	29	9	2
101	Mendelevium, Mv	2	8	18	32	30	9	2
102	Nobelium, No	2	8	18	32	31	9	2
103	Lawrencium, Lw	2	8	18	32	32	9	2

the valence shell

The outermost shell of an atom is called the *valence shell.* The word valence is a Greek word meaning hook; it came into use when an old chemical theory considered that atoms had hooks that held them to other atoms. Since we now know that it is the electrons in the outermost shell that enable atoms to join, the word valence was carried over as the name of the outer shell. Electrons that orbit in the outer shell are also known as *valence electrons.*

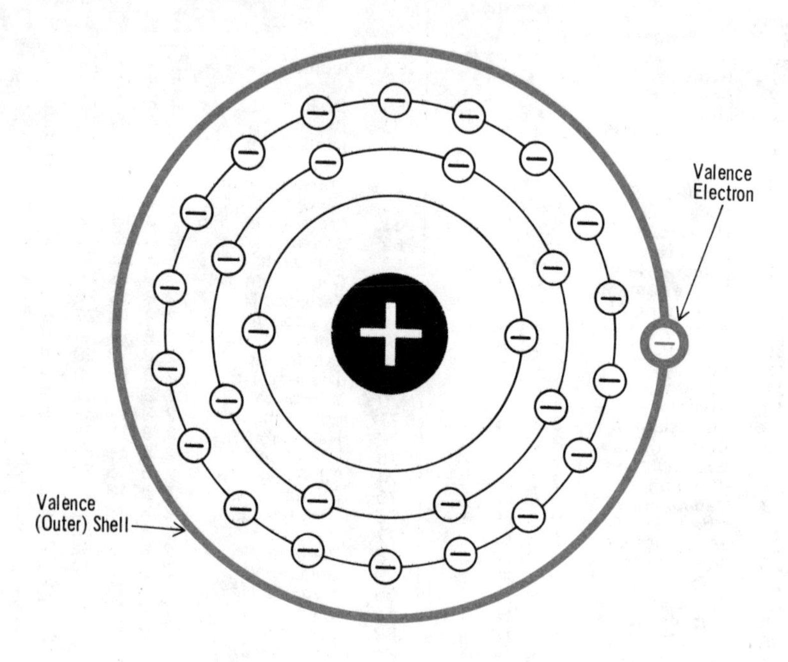

The outer shell is called the VALENCE SHELL, and the electrons in that shell are called VALENCE ELECTRONS

You may have noticed in the discussion on page 4-6, and in the table on page 4-7, that although the third shell can hold up to 18 electrons, it did not have any more than 8 until a fourth shell started. This is also true of the fourth shell. It will not take on any more than 8 until a fifth shell starts, even though the fourth shell can hold up to 32 electrons. This shows that there is another rule: *The outer shell of any atom cannot hold any more than 8 electrons.* This rule is important because it shows which atoms make up good conductors, insulators, or semiconductors, as you will soon learn.

energy levels

Although every electron has the same negative charge, not all electrons have the same *energy level*. Electrons that orbit close to the nucleus have less energy than those that orbit farther away. The farther the electron orbits from the nucleus, the greater the energy it contains. Actually, the energy contained by the electron determines how far it will orbit. Therefore, if we could add energy to an electron in an inner orbit, we can move it out of that orbit to a higher orbit. And, if enough energy is added to a valence electron, it can be moved out of its orbit; and since there is no higher orbit, the electron will be *freed* from its atom.

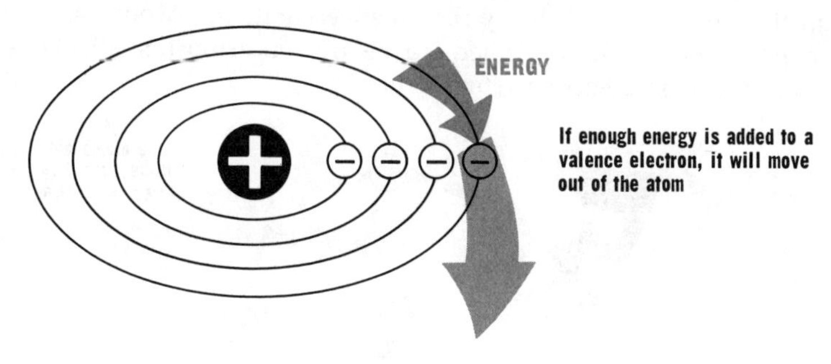

For the sake of simplicity, all the inner electrons are not shown

When energy is applied to an atom, by heat or voltage or by other ways, the shell that first receives the energy is the valence shell. Therefore, valence electrons are the ones most easily removed from an atom. This is easy to understand when you consider that valence electrons are also farthest from the attraction of the nucleus, and so are easier to set free.

Energy is applied to the valence shells, and is distributed amongst the valence electrons

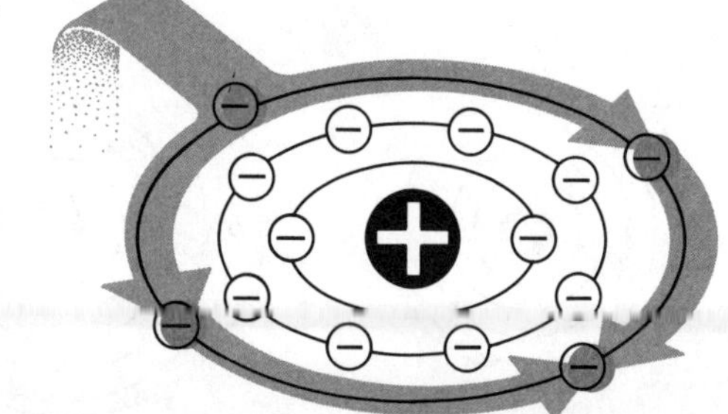

stable and unstable atoms

The tendency of an atom to give up its valence electrons depends on *chemical stability*. When an atom is stable, it resists giving up electrons, and when it is unstable, it tends to give up electrons. The level of stability is determined by the number of valence electrons, because the atom strives to have its outer or valence shell completely filled.

If an atom's valence shell is more than half filled, that atom tends to fill its shell. So, since 8 is the most electrons that can be held in the valence shell, elements with 5 or more valence electrons make good *insulators*, since they tend to take on rather than give up electrons. On the other hand, atoms with less than 4 valence electrons tend to give up their electrons to empty the valence shell; this would allow the next shell, which is already filled, to be the outermost shell. These atoms make the best electrical *conductors*.

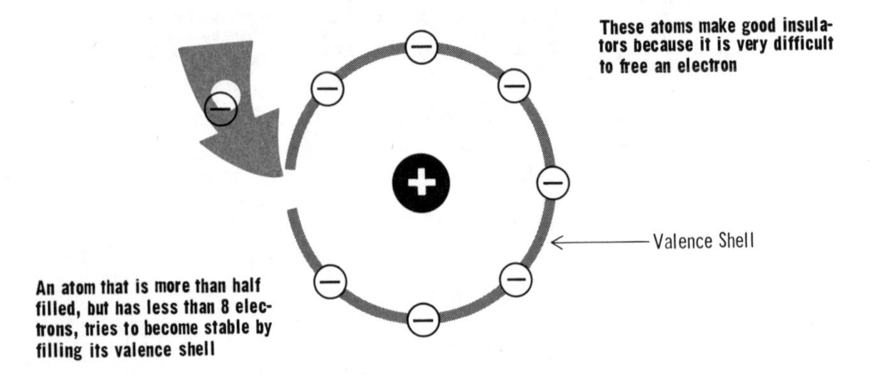

An atom with 8 valence electrons is completely stable, and will resist any sort of activity. These are the *inert gases* (atomic numbers 2, 10, 18, 36, 54, and 86), and are the best insulators. Atoms with only 1 valence electron are the best conductors. As you probably have gathered by now, *semiconductors* have 4 valence electrons, and are neither good conductors nor good insulators.

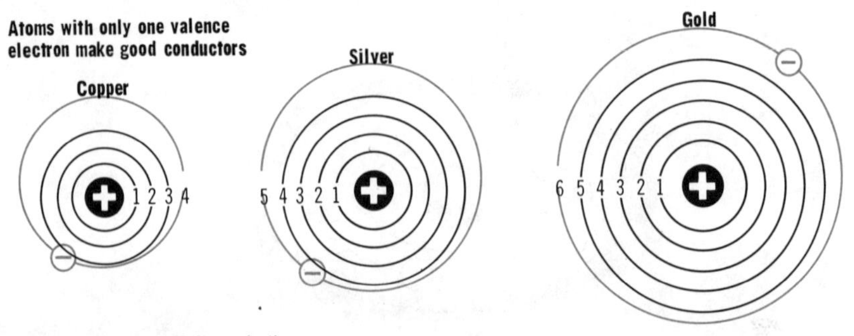

For the sake of simplicity, only the valence electrons are shown

energy band diagrams

Since it is *easy* to force a valence electron from an atom of a good conductor, it takes only *little energy* to do this. And, similarly, since it is *difficult* to free an electron from an atom of an insulator, it takes *more energy* to do this. Semiconductors require less energy than insulators, but more than conductors.

Actually, the energy required to release electrons in the three different types of material falls into a certain band called the *conduction band,* and can be shown on *energy band* diagrams to indicate how easy or difficult it is to free an electron to start electrical conduction. The energy band diagrams on this page show only the energy levels from the valence through the conduction bands. Actually, there are other energy levels, one for each shell down to the nucleus of the atom. But we are only interested in the band beginning with the valence shell.

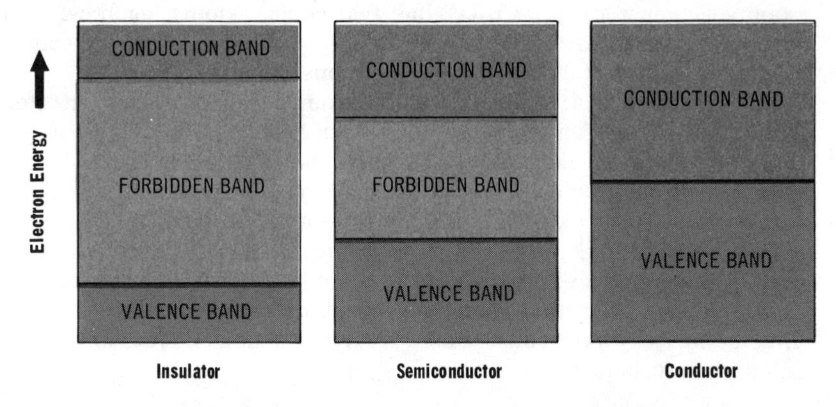

Energy band diagrams show the energy needed to move a valence electron out of the atom to cause conduction

The height of the forbidden band shows the energy needed to free a valence electron

The valence band in each diagram stands for the energy level of the valence electrons; and the conduction band stands for the energy level that must be reached for the valence electrons to be freed from the valence shell. The forbidden band, when it exists, is the energy gap between the other two bands. If only enough energy is added to a valence electron so that its total energy lies in the forbidden band, the valence electron will not be freed; it will stay in the valence shell. You can see, then, that the height of the forbidden band indicates how easy or difficult it is to free a valence electron and start conduction. And, as the diagrams show, insulators have a high forbidden band, semiconductors have a thinner forbidden band, and conductors have no forbidden band. Actually, the valence and conduction bands in good conductors overlap, so that valence electrons in these materials move randomly from one energy level to the next, and continuously move out of the valence shell of one atom into that of another; that is why it is very easy to cause current flow in conductors, as you will learn later.

atomic bonds

Until now, you have studied the characteristics of individual atoms in relation to conductors, semiconductors, and insulators. Actually, we will not use individual atoms in electronics. The atoms must join to form molecules of materials before they can be put to use. And, when these atoms join, their characteristics quite often change because of the chemical nature of the *bond.*

When atoms join to form compounds their characteristics change because the valence shells of the individual atoms *appear* filled. Atoms combine, generally, so that the sum of the valence electrons is 8. In the case of water (H_2O), there are two hydrogen atoms and one oxygen atom in every molecule. Since each hydrogen atom has one valence electron, and the oxygen atom has 6, the molecule has a total of 8. This similar arrangement occurs with most other atomic bonds. For example, copper oxide (Cu_2O) has two copper atoms, each with one valence electron, and one oxygen atom, with six valence electrons; again, there is a total of 8 valence electrons.

Copper oxide clearly illustrates the change brought about in atoms when they form compounds. Pure copper, as you know, is a good conductor because each atom has only one valence electron. But the copper oxide molecule has 8 valence electrons, which makes it stable; as a result, copper oxide is a good insulator. This is true of most compounds. However, keep in mind that the electrical characteristics of many compounds are also affected in other ways, especially by temperature. You will learn more about this later. In any event, you can better understand why molecules tend to become stable if you know more about atomic bonds.

When atoms bond to form molecules, they combine in such a way that the molecule contains 8 valence electrons – a completed outer shell

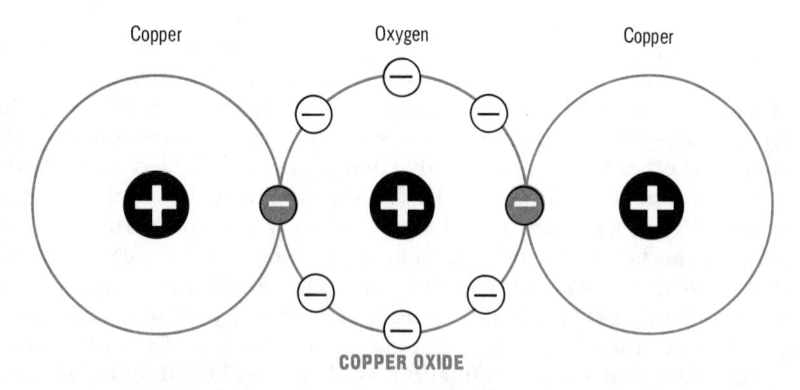

Elements that are normally good conductors, such as copper, can form compounds that are good insulators, such as copper oxide, because the molecule has a completed outer shell

metallic and electrovalent bonds

Atoms are thought to become bonded primarily by *sharing valence electrons.* Electrons can be shared by metallic bonding, electrovalent bonding, and covalent bonding.

Metallic bonding occurs mostly with such good conductors as copper. Each atom has 1 valence electron that follows the conductor energy band diagram shown on page 4-11. There is no forbidden band, and the energy level of the valence electron overlaps into the conduction band. As a result, the valence electron randomly leaves its orbit, but encounters another orbit almost immediately, freeing the valence electron of that atom to repeat the process. This valence electron travel occurs continuously in a random manner so that any one valence electron is always in some orbit, but is not associated with any one particular atom. Therefore, all of the atoms share all of their valence electrons and are bonded.

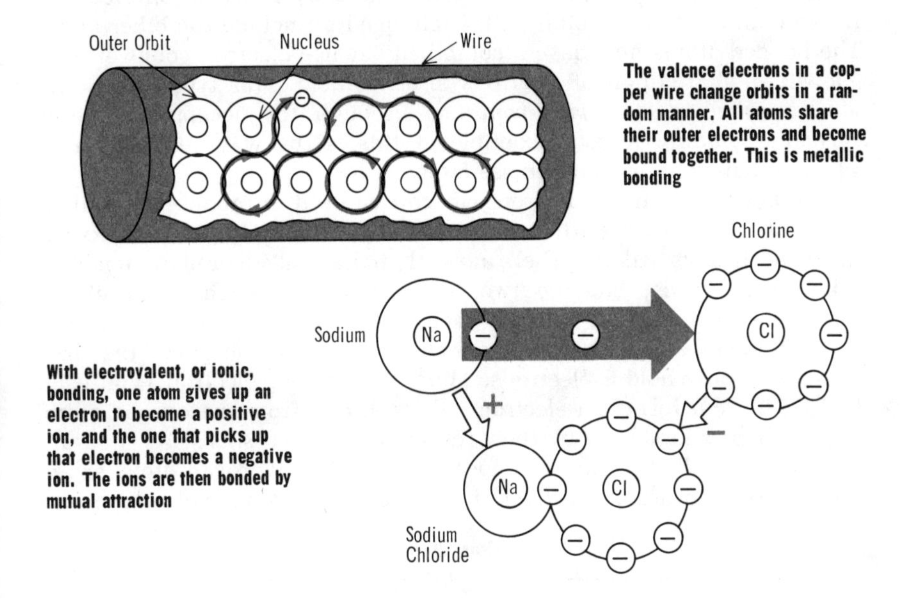

The valence electrons in a copper wire change orbits in a random manner. All atoms share their outer electrons and become bound together. This is metallic bonding

With electrovalent, or ionic, bonding, one atom gives up an electron to become a positive ion, and the one that picks up that electron becomes a negative ion. The ions are then bonded by mutual attraction

Electrovalent bonding takes place when atoms of different elements give up or gain valence electrons to or from each other. When this happens, the atoms then have more or less electrons than protons, and so take on an electrical charge; they become *ions.* Since one atom gives up an electron to the other atom, the one that gives up the electron becomes a positive ion and the one that takes it on becomes a negative ion; and since opposite charges attract, the mutual attraction of the ions forms an atomic bond. Since ionic *attraction* is involved, this is also known as *ionic bonding.* Notice in the illustration for sodium chloride (common salt), that the ions combine to form a molecule with 8 valence electrons.

covalent bonds

You may have noticed that atoms that take part in metallic or electrovalent bonding must have certain valence characteristics. In metallic bonding, the atoms must be good conductors with only one valence electron. And in electrovalent bonding, one atom must be unstable so that it gives up an electron to another atom that actively tries to take on an electron. You might wonder, then, what happens when different atoms that resist either giving up or taking on electrons are combined. They must *share* electrons to be bonded. These atoms share electrons in interesting ways.

When two such atoms meet, *each* atom allows one of its valence electrons to be shared by the other. For example, if we have two atoms each having 4 valence electrons, each will allow one to be shared; so they each keep 3 in their own personal orbit, and 2 move alternately from orbit to orbit. In this way, neither actually gives up on electron. Instead, an electron's orbital path is changed to include the other atom. The bonded atoms now have a combined valence electron count of 8 by allowing a shared pair of electrons to bond them. This type of bonding is also called *electron-pair bonding,* although the accepted name is *covalent bonding.* A good example of this is the way germanium or silicon atoms join, as you will see later.

Another way that an electron pair can be used to produce a covalent bond can be shown with a molecule of water (H_2O). The oxygen molecule has its valence shell more than half filled, and so tends to take on electrons; but the two hydrogen atoms each resist giving up or taking electrons because their valence shells are exactly half filled. Remember that a hydrogen atom has only one shell, the first, which can only hold 2 electrons. Thus, the valence electron from each hydrogen atom forms an electron pair with one from the oxygen atom, and each pair is shared by their respective valence shells. As a result, the shells, of all three atoms appear filled. Each hydrogen atom has 2, and the oxygen atom effectively has 8; also the entire molecule has 8.

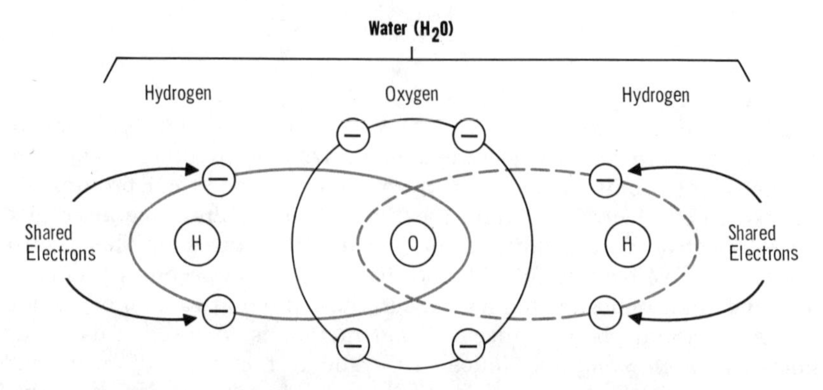

With covalent bonding, a pair of electrons is shared to complete the valence shells of the individual atoms

summary

□ Atoms of the various elements differ from each other only in the number of subatomic particles each contains. □ Protons and neutrons are in the nucleus of the atom, and the electrons revolve in orbits around the nucleus. □ The electrons can orbit in up to seven shells with each shell containing a specific number of electrons. □ The outermost shell is the valence shell and can contain up to 8 valence electrons.

□ An atom strives to fill its outer shell with 8 valence electrons. □ Atoms with 5 or more valence electrons tend to take on electrons, and thus make good insulators. □ Atoms with less than 4 valence electrons tend to give up electrons, and thus make good conductors. □ Semiconductors have half-filled outer shells, and are neither good insulators nor good conductors. □ Insulators and semiconductors have a forbidden band that represents the amount of energy needed to move a valence electron out of the atom to cause conduction.

□ Atoms bond together to fill their outer shells by sharing their valence electrons. □ Electrons are shared by metallic bonding, electrovalent bonding, or covalent bonding. □ Metallic bonding is the random travel of valence electrons from one atom to another to maintain each outer shell filled. □ Electrovalent, or ionic, bonding is the giving up and taking on of valence electrons by atoms of different elements to form positive and negative ions. The positive and negative ions are bonded by the attraction of the unlike electrical charges. □ Covalent bonding is the sharing of a pair of electrons between atoms to complete the valence shells of the individual atoms.

review questions

1. What particles are found in the nucleus of an atom? In the orbits?
2. What is a *shell*? How many electrons can each shell hold?
3. What is the maximum number of valence electrons?
4. How many valence electrons are in an insulator? In a conductor? In a semiconductor?
5. What is the *forbidden band*?
6. Why do atoms bond together? How is bonding accomplished?
7. What is *metallic bonding*? Give an example.
8. How does electrovalent bonding occur?
9. What is another name for electrovalent bonding?
10. Explain covalent bonding and give an example?

semiconductor atomic structure

The bond that you will have to be familiar with in your study of transistors and other semiconductor devices is the covalent, or electron-pair, bond. Germanium and silicon are joined by covalent bonds, and these materials, particularly silicon, are the ones used most in semiconductor electronics.

Because of the nature of covalent bonding, the atoms of a semiconductor distribute themselves in a definite geometric pattern. The position of each atom in relation to its bonded atom becomes important if electrons are to pair off. The structure of the atomic pattern that is produced is known as the *crystal lattice*. Crystalline materials are so called because the basic atomic pattern repeats throughout the molecular structure of the material. Other types of materials contain relatively random atomic patterns.

Covalent bonds cause the atomic structure of semiconductors to repeat the same geometric pattern throughout the material – the crystal lattice

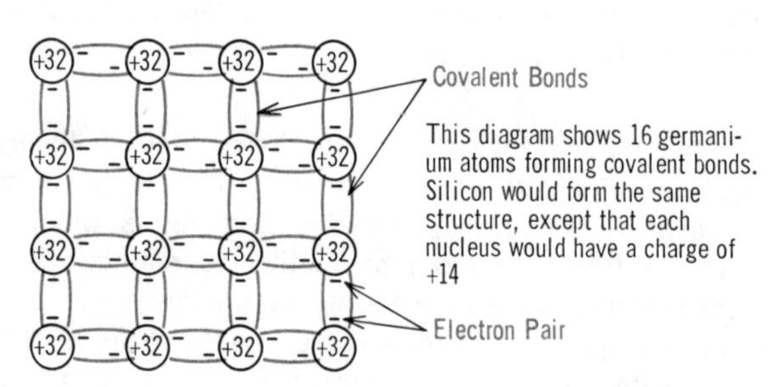

Germanium and silicon have 4 valence electrons each, and the crystalline arrangement allows each valence electron to pair off with an electron of an adjacent atom. Therefore, each atom produces a covalent bond with four adjacent atoms. And each adjacent atom does the same with four others, and so on. This is true of all of the atoms, except, of course, those at the surface of the material, which may not have adjacent atoms. This is why, by the way, the surface of some crystalline materials often does not have the same electrical characteristics as the interior of the material.

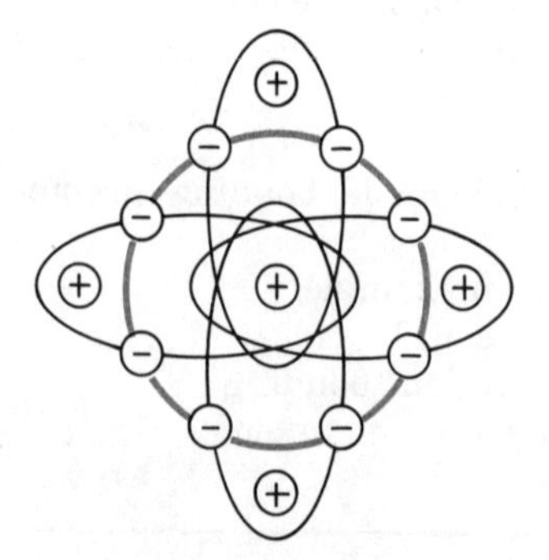

Germanium and silicon atoms form an electron-pair bond with four adjacent atoms. Each adjacent atom then repeats this with four other atoms to produce a crystal lattice

free electrons and holes

A pure semiconductor material allows each of its atoms to "see" 8 valence electrons. Thus, each atom tends to be stable and the entire material acts as an insulator. There are no "free" electrons, which is the term applied to those valence electrons that have an energy level in or near the conduction band so that they can be set free.

In practical use, this is not completely true because, even at room temperature, enough heat energy is available to allow some semiconductor valence electrons to raise their energy level to leave their valence shells. Such electrons are freed and can wander randomly from the valence shell of one atom to another. Now, since the electron-pair bonds between the individual atoms made use of *all* the valence electrons of the atoms, the existence of any *free* electron causes an electron-pair bond to be broken. In such an atom, then, there is a space that should have an electron that does not. This space is called a *hole*. There was a valence electron there, but it was set free by *thermal agitation*.

From a practical viewpoint, therefore, a pure semiconductor, or any compound, for that matter, does have some free electrons that can take part in current flow. And, in addition, semiconductors have holes in the covalent bonds that will readily accept electrons. There is another point to consider. Since the holes will readily accept electrons, and the energy level of various valence electrons is being raised by thermal agitation, many electrons will have their energy level increased to the point where, although they might not be free to wander randomly, they can jump from one shell to fill a hole in the next. When this happens, the hole, effectively, jumps to the covalent bond where that electron came from. That hole, in turn, might be filled by another valence electron, and so the hole "appears" to move again. As a result, a pure semiconductor in practical use has some free electrons and holes moving about in a random way.

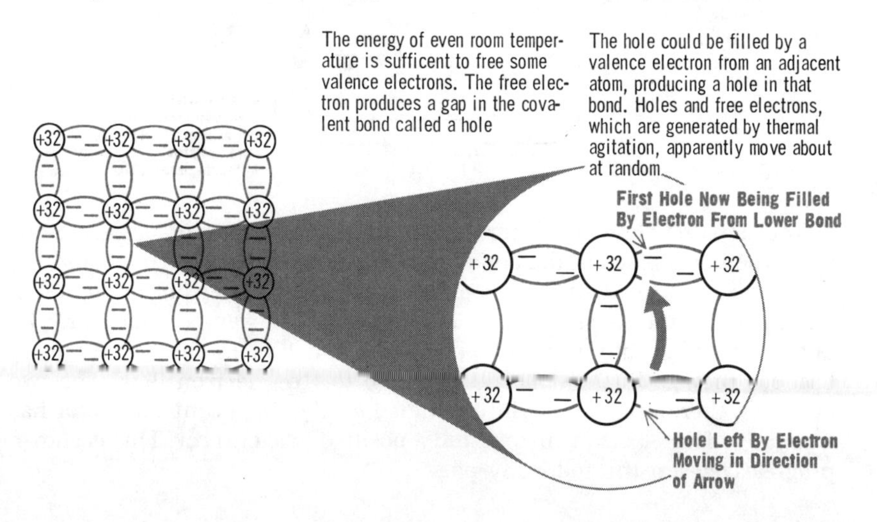

current flow in a pure semiconductor

Even though a pure semiconductor tends to produce stable atoms, it still *conducts* current with some free electrons the same way a good conductor does. However, a pure semiconductor only has a few free electrons, so only a *slight current* flows. The semiconductor, then, has a *high resistance*. Keep in mind, though, that since free electrons are released by thermal agitation, more current will flow at higher temperatures.

Semiconductor current does differ from that in a normal conductor because of holes in the covalent bonds. In a good conductor all of the valence electrons wander as free electrons to make up current flow; but in a semiconductor, only those freed from the electron-pair bonds can take part in electron current. The holes, though, allow the other valence electrons to move about because the holes continually try to be filled. Those valence electrons do not have to reach the energy level of the conduction band. Their levels need only be in the forbidden or even in the valence band because they only move to an adjacent atom, and the attraction of the hole provides the additional force for them to move. The gap in the covalent bond acts just like a *positive* charge, and so the hole is considered as such.

If a voltage were applied across the semiconductor, the resulting attraction would draw the valence electrons toward the positive potential. This, in turn, would make the holes appear to drift toward the negative potential. The free electrons, of course, would go toward the positive voltage, as in any conductor.

If voltage is applied across a semiconductor, free electrons drift toward the positive side as in any ordinary conductor

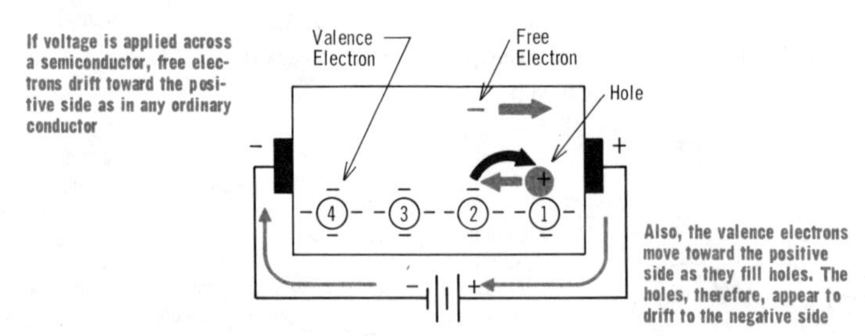

Also, the valence electrons move toward the positive side as they fill holes. The holes, therefore, appear to drift to the negative side

You can see then that there are actually two types of current flow. One type takes place in the conduction band with free electrons, and the other takes place in the valence band with valence electrons or holes. Even though the holes are only an "apparent" movement, they are the ones that are usually used to describe current flow in the valence band because they go in the opposite direction to free electron flow. The two currents, therefore, can be differentiated easily. So, a semiconductor has a negative free electron current and a positive hole current. This is shown progressively on the following page.

semiconductor current

This shows four atoms. When voltage is first applied to the semiconductor, there are four atoms, and a hole in the bond of atom 1. There is one free electron.

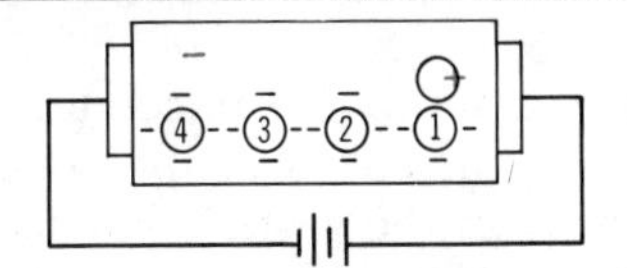

The positive voltage of the battery attracts the free electron to its side, starting electron current. The positive voltage also attracts a valence electron from atom 2, causing it to fill the hole in atom 1. The hole now exists in atom 2.

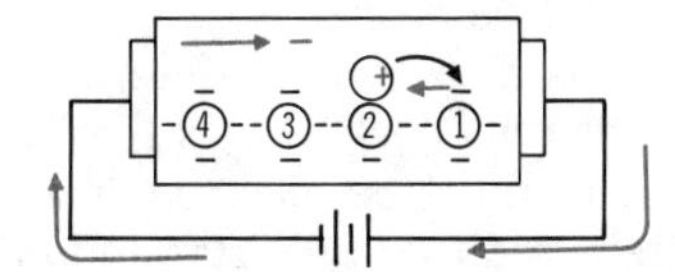

The free electron continues to move toward the positive side. Now a valence electron is attracted from atom 3 to fill the hole in atom 2. The hole is now in atom 3.

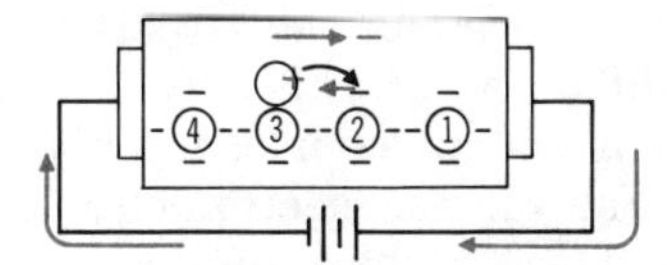

The free electron reaches the positive side. And, a valence electron leaves atom 4 to fill the hole in atom 3. The hole moves to atom 4 at the negative terminal.

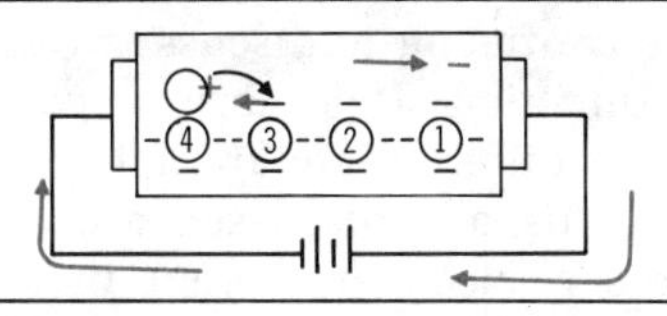

The free electron is attracted from the semiconductor to take part in the wire current flow. At the same time, a free electron from the battery enters the semiconductor at the negative side to replace the one that left, so that the free electron current can continue to flow. In addition, the hole in atom 4 attracts a free electron from the wire and is filled. Now, this free electron went from a high energy level, the conduction band, to a lower level, the valence band. To do this it gave up energy. The energy is transferred to the other valence electrons, and is picked up by the valence electron in atom 1, since it is closest to the positive attraction of the battery. The valence electron in atom 1 then is raised to the conduction band. It is freed and leaves the semiconductor to take part in the electron current flow in the wire. Now, a hole is back in atom 1 to replace the one that was lost in atom 4.

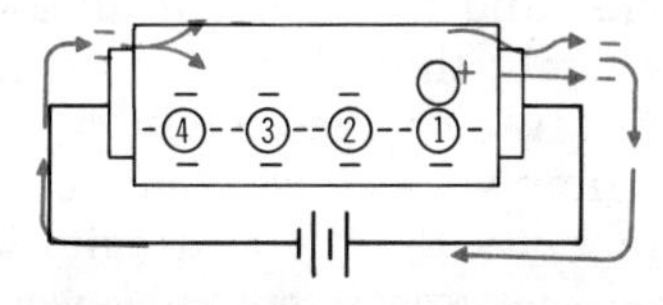

Electron current flow continues because a free electron enters the semiconductor on the negative side for each one that leaves on the positive side. Also, for each hole that is lost, or filled, at the negative side, a new one is created at the positive side. The holes do not leave or enter the semiconductor; they cannot because there are no covalent bonds or holes in the wire. Instead, the hole current is maintained by electrons entering or leaving the semiconductor.

The current flow in the wire is in the same direction for the free electron flow and hole flow. Therefore, both of these currents in the semiconductor are added in the wire.

free electrons, valence electrons, and holes

Current flow in a semiconductor takes part at two different energy levels: one current is in the conduction band with free electrons; and the other is in the valence band with valence electrons, or holes. In an ordinary conductor, current flow takes place only in the conduction band with free electrons.

Remember, the only difference between a free electron and a valence electron is the energy levels they contain. Both are found in the valence shell. Only the free electron is not part of a covalent bond and is so easily released that it drifts randomly. The free electron does *not* wander free of any atom, but it is usually shown that way for convenience.

The valence electron is part of an electron-pair bond and is released with difficulty, and usually only when it is in the vicinity of a hole. Hole current, of course, is only the apparent movement of the holes due to the valence electron current. However, hole current is universally accepted as the current that occurs in the valence band, because it can be confusing to discuss two different electron currents going in the same direction. Hole current goes in the opposite direction, and so is easier to differentiate from free electron current. As a result, the two currents usually discussed are electron current and hole current. But remember, electron current in a semiconductor always refers to free electrons, and hole current always refers to valence electrons.

You might think that there is no reason to differentiate the two currents because in the examples you studied, both electron current and hole current were equal. You recall that the holes and free electrons were both produced by thermal agitation; and a hole was created for each electron that was freed. However, this is only true for a pure semiconductor. You will learn next that semiconductors can be manufactured to have more free electrons than holes, and vice versa. Also, when you think about it, you can see that if a low voltage is applied across the semiconductor, it could move the free electrons, which have a high energy level, but not as many valence electrons, which are at a lower energy level. So, the ratio of the two currents that flow depends on the voltage applied. Current flow in a semiconductor, then, tends to be *nonlinear* at the lower voltages, but could become linear at the higher voltages. Current flow in a conductor is linear. Keep in mind that because of this, it is difficult to use Ohm's Law with semiconductors.

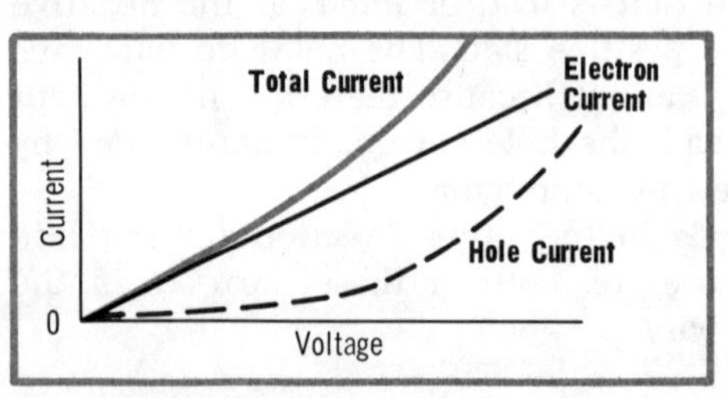

Since hole current depends on low-energy valence electrons, very little might flow at low voltages. When the voltage is sufficiently high, hole current could flow linearly. Free electron current does tend to flow linearly, but the total current shows that the semiconductor is a nonlinear device

current carriers and doping

You have learned that semiconductors have two types of current flow: electron current and hole current. Electron current flow is not actually the "flow" of an electron through a material, but it is the electrical impulse that one electron imparts to another when it starts to move. The electrons do not travel straight through the material to carry the current. Each electron may only move slightly in the direction of the current, but its transfer of energy to other electrons is what comprises the flow. This is easier to visualize with the valence electrons or holes in the illustrations on page 4-19. It took the movement of four *different* valence electrons to get one to leave the material; the holes, too, moved in the same manner. Four different holes had to be created or lost to cause the apparent motion of one hole. Holes and electrons, then, are not the current in themselves. But they are the *carriers* of the current, and this is how they are referred to. Free electrons are *negative current carriers,* and holes are *positive current carriers.*

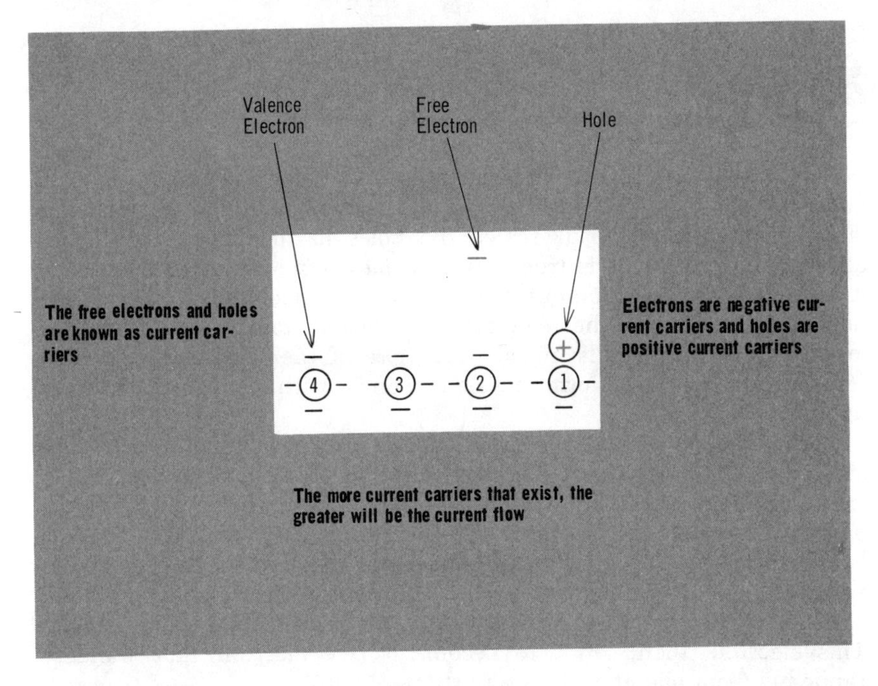

Although the carriers are not the currents in themselves, they do determine how much current will flow. The number of carriers that moves determines how much current energy will be transferred through the semiconductor. A pure semiconductor has only a few current carriers, particularly at the lower temperatures, so very little current can flow. However, semiconductor materials can be made to have more carriers and to provide more current flow. This is done by a process called *doping.*

Atoms that have 5 valence electrons are PENTAVALENT atoms

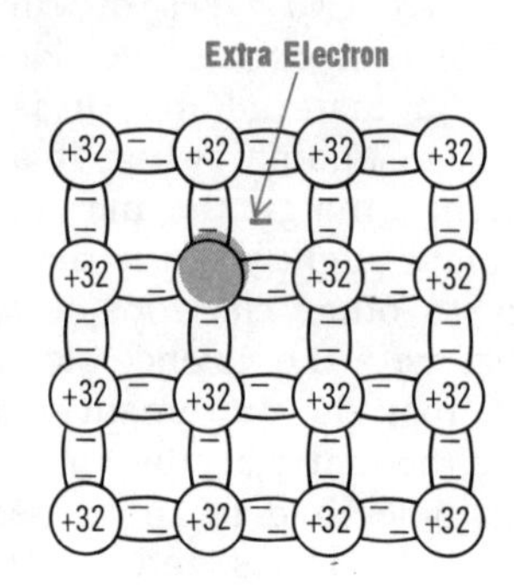

If the impurity atom has 5 valence electrons, one will not be able to take part in an electron-pair bond. That electron will become a free electron

Atoms that have 3 valence electrons are TRIVALENT atoms

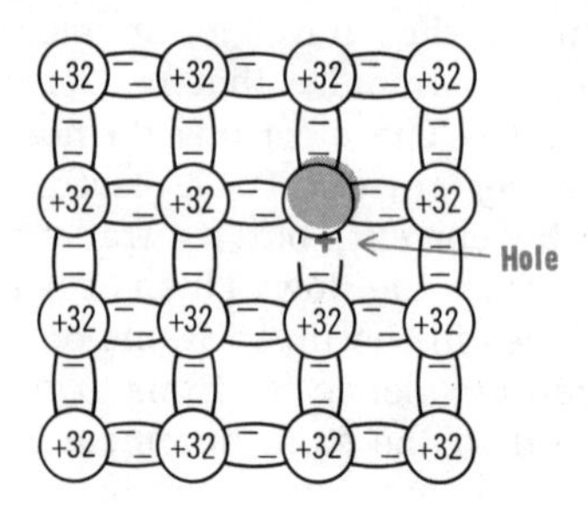

If the impurity atom only has 3 valence electrons, one covalent bond will have a hole in it, which can act as a current carrier

doping

You have learned that free electrons and holes are produced in a pure semiconductor because of thermal agitation, and that since the electrons that were freed created the holes in the bonds, they were equal in number. This is true only in a *pure* semiconductor. However, it is very difficult to manufacture a pure semiconductor. There are usually *impurities* in the material. This merely means that there are atoms of other materials mixed with those of the semiconductor. The impurity atoms also form covalent bonds with the semiconductor atoms.

The existence of these impurities in the semiconductor can considerably change the characteristics of the material, depending on how many impurity atoms are present. For example, if the impurity atoms have 5 valence electrons, only 4 are needed to take part in the covalent bonds. Therefore, there is an extra electron that is loosely held in the valence shell. Since the atoms "see" 8 valence electrons without the extra one, they try to eliminate the extra electron to become stable. This electron, then, tends to become a free electron that wanders randomly from one atom to the next.

If, on the other hand, the impurity atoms only have 3 valence electrons, they would allow gaps, or holes, to exist in their covalent bonds. In any event, you can see that certain types of impurity atoms in the semiconductor increase the number of current carriers.

When the semiconductor material is being manufactured, the addition of impurities can be *controlled* to give the semiconductor any desired electrical characteristic. The addition of impurities is known as *doping*.

penta- and trivalent impurities

Semiconductor materials can be given electrical characteristics to make them more useful by doping them with impurity atoms that either cause an excess of free electrons or holes. The impurity atoms that are added have either 3 or 5 valence electrons. The impurity atom that has only 3 valence electrons is called a *tri*valent impurity. Some typical trivalent impurities are indium, gallium, and boron. The impurity atom that has 5 valence electrons is called a *penta*valent impurity. Some examples of these are arsenic, phosphorus, and antimony.

When a trivalent atom forms a bond with the semiconductor atoms, it only has 3 valence electrons that can pair off with 4 from the adjacent semiconductor atoms. There is, therefore, a gap left in one covalent bond of the trivalent atom. This hole in the bond acts as a positive charge that tends to accept electrons to fill the gap. As a result, these impurities are called *acceptor impurities*. They have atoms with holes that accept electrons.

The pentavalent impurity atom has an electron left over after the covalent bonds are made. It is called a *donor* atom, since it donates a free electron to the semiconductor.

To summarize, trivalent, or acceptor, impurities provide excess holes, or positive current carriers; and pentavalent impurities provide excess free electrons, or negative current carriers.

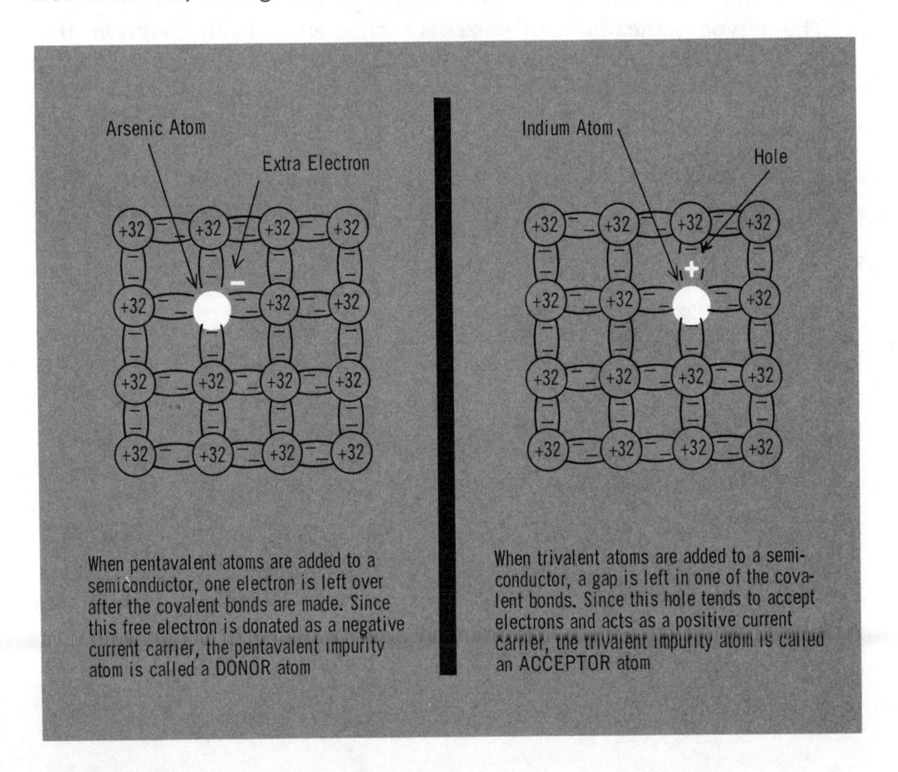

When pentavalent atoms are added to a semiconductor, one electron is left over after the covalent bonds are made. Since this free electron is donated as a negative current carrier, the pentavalent impurity atom is called a DONOR atom

When trivalent atoms are added to a semiconductor, a gap is left in one of the covalent bonds. Since this hole tends to accept electrons and acts as a positive current carrier, the trivalent impurity atom is called an ACCEPTOR atom

p- and n-type semiconductors

When donor atoms are added to a semiconductor, the extra free electrons give the semiconductor a greater number of free electrons than it would normally have. And, unlike the electrons that are freed because of thermal agitation, donor electrons do not produce holes. As a result, the current carriers in a semiconductor doped with pentavalent impurities are primarily negative. Such a semiconductor, as a result, is called an *n-type* semiconductor.

Since n-type semiconductors have extra free electrons, and pure semiconductors do not, the energy band diagram for a doped semiconductor is slightly different from that of a pure semiconductor. In effect, another energy level exists: a level for the donor electron, which is closer to the conductor band. The forbidden band for the donor electron is much narrower than the forbidden band for the valence electron; so, you can see that it is much easier to cause electron flow in an n-type semiconductor.

When acceptor atoms are added to a semiconductor material, more holes are produced than there would have been from thermal agitation alone. And, unlike the holes that are produced from thermal agitation, electrons did not have to be freed to cause them. As a result, the current carriers in a semiconductor doped with trivalent impurities are primarily positive. Such a semiconductor is called a *p-type* semiconductor.

The p-type semiconductor also has an energy band diagram that differs from that of the pure semiconductor. Since there is an extra number of holes, which tend to attract electrons, they aid in starting current flow. As a result, the acceptor energy level is also somewhat higher than that of the valence band. However, it is not as high as the donor level. P-type semiconductors will conduct current easier than pure semiconductors, but not quite as easy as n-type semiconductors.

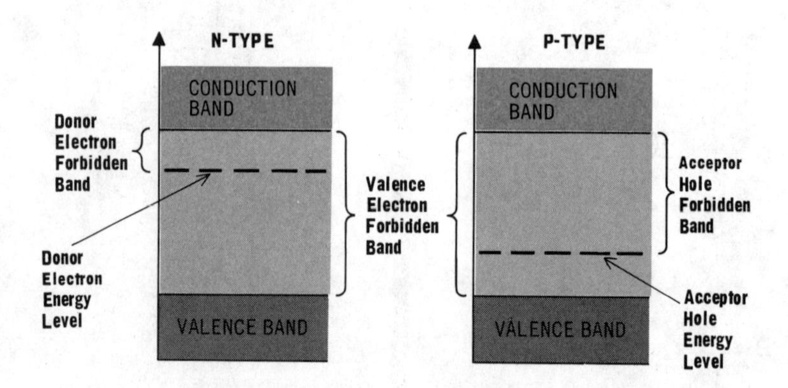

Excess free electrons in n-type semiconductors produce a donor energy level close to the conduction band. Excess holes in p-type semiconductors introduce an acceptor energy level higher than the valence level, but not as high as the donor level. N-type semiconductors, then, conduct current easier than p-type semiconductors

Bookmark

HOWARD W. SAMS & COMPANY

DEAR VALUED CUSTOMER:

Howard W. Sams & Company is dedicated to bringing you timely and authoritative books for your personal and professional library. Our goal is to provide you with excellent technical books written by the most qualified authors. You can assist us in this endeavor by checking the box next to your particular areas of interest.

We appreciate your comments and will use the information to provide you with a more comprehensive selection of titles.

Thank you,

Vice President, Book Publishing
Howard W. Sams & Company

COMPUTER TITLES:

Hardware
- ☐ Apple 140 ☐ Macintosh 101
- ☐ Commodore 110
- ☐ IBM & Compatibles 114

Business Applications
- ☐ Word Processing J01
- ☐ Data Base J04
- ☐ Spreadsheets J02

Operating Systems
- ☐ MS-DOS K05 ☐ OS/2 K10
- ☐ CP/M K01 ☐ UNIX K03

Programming Languages
- ☐ C L03 ☐ Pascal L05
- ☐ Prolog L12 ☐ Assembly L01
- ☐ BASIC L02 ☐ HyperTalk L14

Troubleshooting & Repair
- ☐ Computers S05
- ☐ Peripherals S10

Other
- ☐ Communications/Networking M03
- ☐ AI/Expert Systems T18

ELECTRONICS TITLES:
- ☐ Amateur Radio T01
- ☐ Audio T03
- ☐ Basic Electronics T20
- ☐ Basic Electricity T21
- ☐ Electronics Design T12
- ☐ Electronics Projects T04
- ☐ Satellites T09
- ☐ Instrumentation T05
- ☐ Digital Electronics T11

Troubleshooting & Repair
- ☐ Audio S11 ☐ Television S04
- ☐ VCR S01 ☐ Compact Disc S02
- ☐ Automotive S06
- ☐ Microwave Oven S03

Other interests or comments: ______________________________
____
____

Name ______________________________
Title ______________________________
Company ______________________________
Address ______________________________
City ______________________________
State/Zip ______________________________
Daytime Telephone No. ______________________________

A Division of Macmillan, Inc.
[illegible]
Indianapolis, Indiana 46268

45957

Bookmark

HOWARD W. SAMS
& COMPANY

NO POSTAGE
NECESSARY
IF MAILED
IN THE
UNITED STATES

BUSINESS REPLY CARD

FIRST CLASS PERMIT NO. 1076 INDIANAPOLIS, IND.

POSTAGE WILL BE PAID BY ADDRESSEE

HOWARD W. SAMS & CO.
ATTN: Public Relations Department
P.O. BOX 7092
Indianapolis, IN 46209-9921

majority and minority carriers

Although an n-type semiconductor conducts more easily than a p-type, this does not mean that one will conduct more than the other for a given voltage. This is because the current flow in each depends on the number of extra carriers that exist, or how much they were doped. If the number of holes in the p-type semiconductor is equal to the number of free electrons in the n-type, more current will flow in the n-type for a given voltage because the donor (free) electrons have a higher energy level than the acceptor holes. But, if the p-type were doped so that it had many more holes, it could conduct more current than the n-type. In effect, the donor and acceptor energy level determines how easy it is to move one electron or hole carrier, but the *number* of carriers moved determines how much current will flow.

The carriers we have been studying are the *main* carriers, or *majority* carriers as they are called, because doping produces an excess of them: free electrons in the n-type, and holes in the p-type. However, there is another carrier: thermal agitation frees some valence electrons, producing an equal number of holes. The freed valence electrons in the n-type merely join the extra free electrons put there by doping to become majority carriers. The holes, though, which effectively flow in the opposite direction, become *minority carriers*. When a voltage is applied to an n-type semiconductor, a relatively heavy free electron majority current flows in one direction, while a smaller hole minority current flows in the opposite direction.

In a p-type semiconductor, the holes produced by thermal agitation merely join the excess holes produced by doping to become majority carriers. The freed electrons, though, because of thermal agitation are much fewer in number, and become minority carriers. So, when a voltage is applied to a p-type semiconductor, a relatively heavy hole majority current flows in one direction while a much smaller free electron minority current flows in the other direction. With both p- and n-types, what is a majority carrier in one is a minority carrier in the other.

Another way to look at the carriers is that the unbonded, or free, electrons traveling from valence shell to valence shell form one current carrier; and bonded, or valence, electrons traveling from hole to hole form the other current carrier.

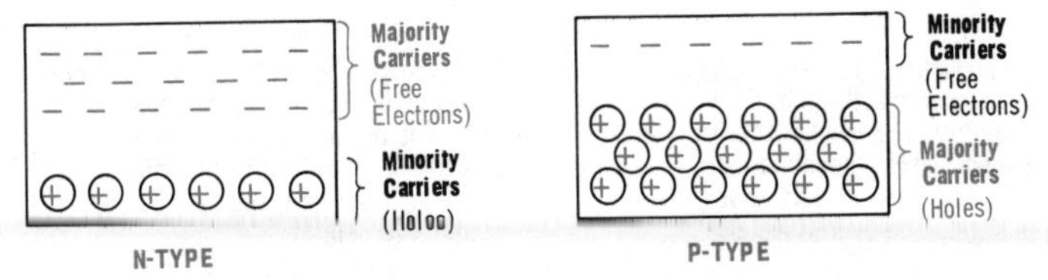

Both p- and n-type semiconductors have two kinds of carriers: majority and minority carriers. The majority and minority carriers flow in opposite directions for any given voltage polarity. In any one type of semiconductor, majority current is much greater than minority current

p and n electrical charges

The basic atom, you recall, normally has the same number of electrons and protons, so that it is neutral. However, atoms go through somewhat of a change when they are part of the crystal lattice of a semiconductor. For example, in the n-type semiconductor, the pentavalent impurity atom uses 4 of its valence electrons to form covalent bonds with its neighboring atoms, and then frees its fifth valence electron. The atom, then, having lost one negative electrical charge, becomes a *positive ion*. And this is true for all other atoms that give up electrons. However, for each positive ion that exists, there is a negative free electron in the semiconductor. So, the overall semiconductor is neutral.

In a p-type semiconductor, the trivalent impurity atom lacks one valence electron, and so causes a hole when the covalent bonds are formed. Normally, because of thermal agitation, these holes are generally filled by valence electrons from the semiconductor atoms, causing the holes to appear in those atoms. During normal atom activity, valence electrons continue to jump their bonds to fill adjacent holes, so that the holes tend to move in a random manner. Since there are more semiconductor atoms than impurity atoms, you will find at any one instant that the hole will exist in the bonds of the semiconductor atoms; and the holes in the impurity atoms will be filled. Since the impurity atoms had to take on extra electrons to completely fill their bonds, they become *negative* ions. However, for each negative ion that exists, there is a positive hole. Another way to look at this is that any semiconductor atom that has a hole had to give up a valence electron, and so these atoms become *positive ions*. In any case, the p-type semiconductor has an equal number of positive and negative charges, and so the overall semiconductor is neutral.

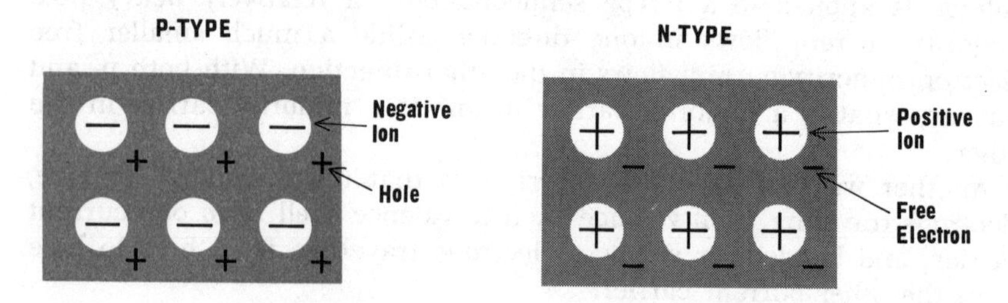

In p-type semiconductors, the semiconductor atoms give up valence electrons to fill the holes in the impurity atom bonds. The impurity atoms, then, become negative ions; but for each of these there is a positive hole in the material. The overall material is neutral

In n-type semiconductors, the impurity atoms give up their excess valence electron. The impurity atoms, then, become positive ions; but for each of these, there is a free negative electron. The overall material is neutral

thermistor, varistor, resistor

Since a semiconductor can be controlled during the doping process, it can be given any desired resistance characteristics. But because of the cost, ordinary *resistors* are not usually made this way. This process is used, though, in integrated circuits, discussed later. By themselves, then, p- and n-type semiconductors find limited use. But, because they have nonlinear electrical characteristics and are affected by temperature, they can be applied in *voltage regulating* circuits.

A doped semiconductor, you recall, has a majority current and a minority current, and the minority current might only become significant at the higher voltages. Therefore, the overall current of a properly doped semiconductor will increase even more as the voltage applied to it is increased than would the current in an ordinary resistor that follows Ohm's Law. As a result, such a semiconductor is said to have a resistance that goes down, or is *inversely proportional*, with *voltage*. One such device is called a *varistor*, and the circuit for using it to regulate voltage is shown.

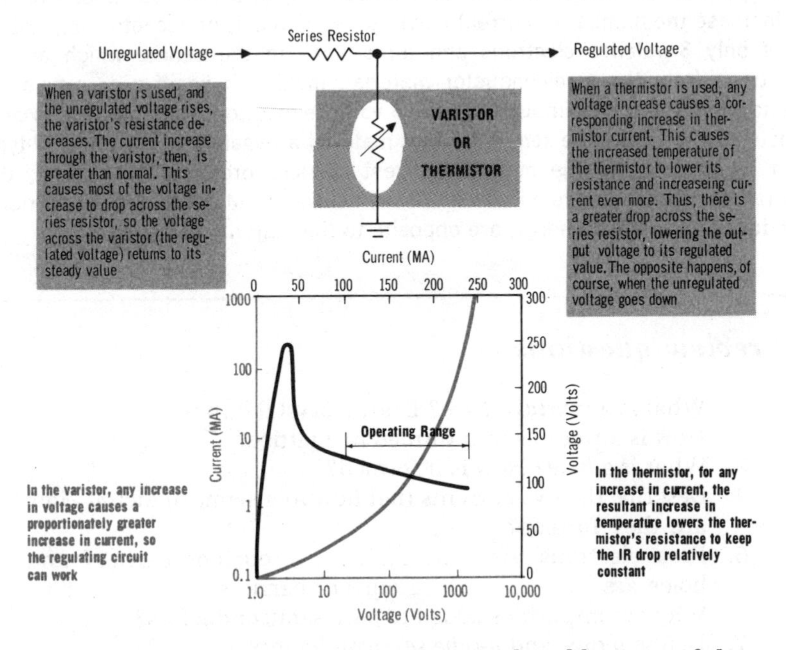

Remember also that minority carriers are produced because of thermal agitation. A semiconductor can be made so that minority carriers will be more easily produced as the temperature of the semiconductor goes up. Such a device, then, is said to have a resistance that goes down, or is *inversely proportional*, with *temperature*. One such device is called a *thermistor*, and can also be used to regulate voltage as shown.

summary

□ Breaking a covalent bond produces a free electron, which moves about the lattice in a random manner. □ The free electron, in leaving a covalent bond, produces a gap in that bond called a hole. □ A valence electron in the lattice moves to fill a hole, forming a new hole in the lattice at a different covalent bond.

□ Two currents flow in a pure semiconductor: a free electron current in the conduction band, and a valence electron current in the valence band. □ The free electron current consists of valence electrons freed from their covalent bond, and flows towards a positive applied potential. □ The valence electron current consists of valence electrons having insufficient energy to break free of the valence band. This current produces the hole current that flows in the opposite direction to the free electron current. □ The free electrons and the holes are not the current in themselves, but act only as carriers of the current. □ Free electrons act as negative current carriers and holes act as positive current carriers.

□ Impurities are added to pure semiconductors, in a process called doping, to increase the number of current carriers. □ Trivalent, or acceptor, impurities have only 3 valence electrons and form holes in the bonds, which accept electrons from the semiconductor material, making it positive or p-type. □ Pentavalent, or donor, impurities have 5 valence electrons, and produce excess electrons that make the semiconductor material a negative or n-type. □ P-type and n-type refer to the majority current carriers produced by doping; the overall semiconductor remains electrically neutral. □ Minority current carriers, produced by thermal energy, are opposite to the majority carriers.

review questions

1. What is a *crystal lattice*? Draw a crystal lattice.
2. How is a free electron formed in a lattice?
3. What is a *hole*? How is it formed?
4. Describe the two currents that flow in a semiconductor. How are they formed?
5. Free electrons are ______________ current carriers and holes are ______________ current carriers.
6. Why are impurities added to pure semiconductors?
7. Define *p-type* and *n-type semiconductors*.
8. Why is an n-type semiconductor material electrically neutral although it possesses an excess of free electrons.
9. How are minority carriers formed?
10. What is a *varistor*? A *thermistor*?

the p-n diode

When p- and n-type semiconductors are combined as a p-n unit, a number of new characteristics are produced, which make the newly formed semiconductor useful. In particular, because each half of a p-n unit has opposite majority and minority carriers, the resistance of the unit to current going in one direction is much higher than the resistance to current going in the other direction. Then, such a unit acts the same as an electron-tube diode, and can rectify a-c currents. It is therefore called a *p-n* or *semiconductor diode*.

When p and n sections are combined, the p sections are not merely pressed against the n sections. Semiconductors are generally "grown" during manufacture. The original method of growing semiconductors was to dip and withdraw a *seed* from a crucible of molten semiconductor material, usually germanium or silicon, and then have the molten material cool and solidify on the seed. As the seed was continually withdrawn, more and more material solidified and accumulated, and so the crystal "grew."

A p-n diode is made with one section that has an excess of hole carriers, and another section that has an excess of free electron carriers

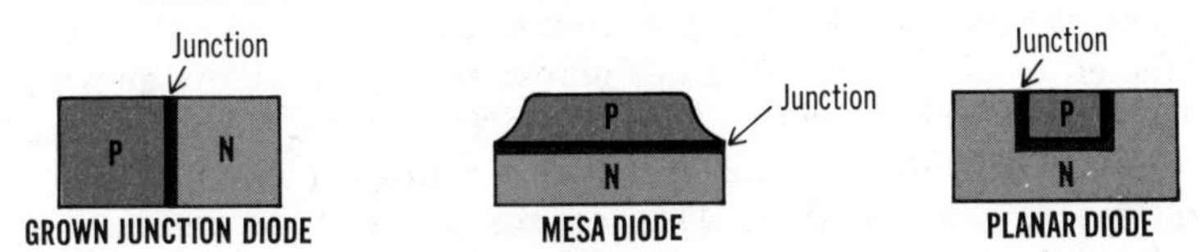

Because of these opposite carriers, the p-n diode conducts more in one direction than in the other

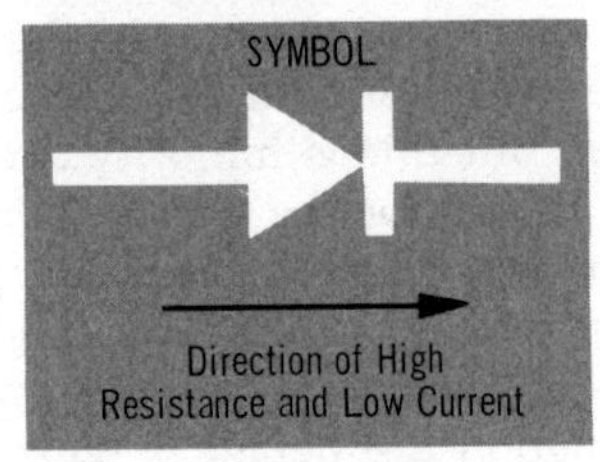

The arrowhead represents the p section, and indicates the direction of the hole flow, or the high-resistance direction

The low-resistance or high-current direction of free electron flow is opposite the arrow

The molten semiconductor material was first given the proper impurities to produce, for example, a p-type, and then when enough was grown, other impurities were added to make the next growth the n-type. So, the p-n diode was actually one unit, with a section called the *junction* separating the opposite segments. This is known as the *grown junction* method of manufacturing a p-n diode. Other methods are the alloy junction, diffusion process, drift technique, meltback methods, mesa, planar, etc. The mesa and planar techniques, which are the two most popular, are discussed later.

the depletion region

P- and n-type semiconductors have two types of electrical charges: the ionic charge the atom takes on when it gives up or receives an electron; and the second charge contained by the current carriers. The current carriers are called *mobile charges*. You know that the two charges are equal and opposite, so that the semiconductor is neutral. However, you recall that the covalent bonds produced a repeating crystal lattice structure, consisting of an atom joined to four others. At the surface of the material, however, this was not the case, since there are not any atoms to complete the structure.

As a result, the opposite charges in a semiconductor do not completely cancel, and the mobile charges at the edges and at the junction are easily affected by external charges. In a circuit, there is usually a *surface* current that is relatively independent of the internal current; in some critical applications, it must be considered. Also, because the mobile charges are easily removed with an applied potential, a *surface barrier* may be produced around the semiconductor to inhibit current flow in and out of the device. This characteristic is the one used by a device known as the *surface barrier transistor,* described later.

A similar effect becomes evident in a p-n diode, after the diode is made, because the mobile charges around the junction in each section are attracted by the mobile charges across the junction. As shown, the free electrons in the n section and the holes in the p section around the junction drift toward the junction because of the attraction of the charges to one another. This causes stronger local charges on each side of the junction. Since the free electrons in the n section have a higher energy than the holes, the free electrons are more mobile. They are attracted by the accumulated holes, and drift across the junction to fill the holes. Sometimes a valence electron in the n section will jump its bond to fill a hole in the p section, and this will cause that hole to move across into the n section. But this is minority carrier action, and occurs in only small numbers, and so for the most part can be ignored.

In any event, since the free electrons crossed the junction to fill holes, these majority carriers on either side of the junction were "lost." After the p-n diode is manufactured, then, the majority carriers in the immediate vicinity of the junction are depleted. This region of the diode is therefore called the *depletion region.*

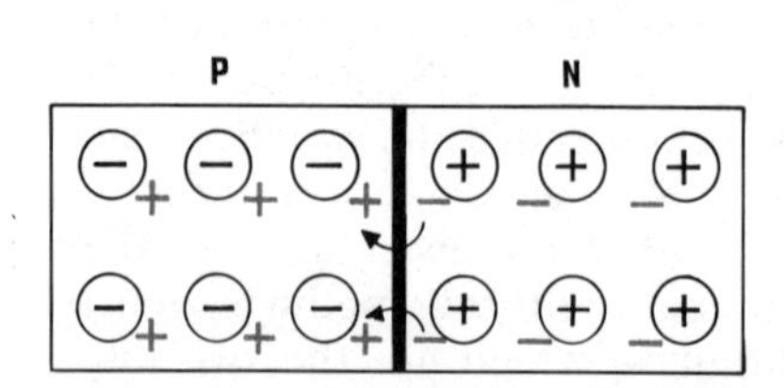

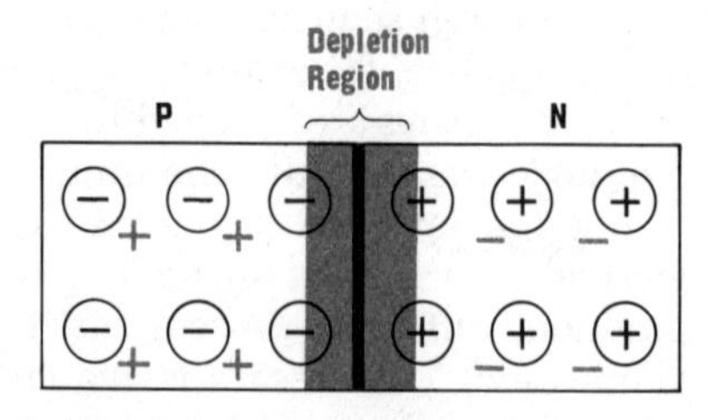

Immediately after the p-n diode is made, a depletion region is produced in the area around the junction

the potential barrier

Before the depletion region was formed, the opposite charges in each section of the p-n diode were equal, so that both sections were electrically neutral. But, after the depletion region was formed, the n section gave up free electrons, or negative charges, and the p section lost holes, or positive charges, as they were filled. As a result, the positive ionic charges in the n section outnumber the negative free electrons, and the n section takes a positive charge.

In the p section, the opposite is true; the p section takes on a negative charge. As more and more free electrons cross the junction to fill holes, these charges become greater and greater. This continues until the overall negative charge of the p section becomes sufficient to repel the free electrons in the n section, keeping them from crossing the junction. At the same time, the overall positive charge built up in the n section attracts the free electrons on their side of the junction. This electrical charge buildup stops the *electron-hole combinations* at the junction, and limits the size of the depletion region.

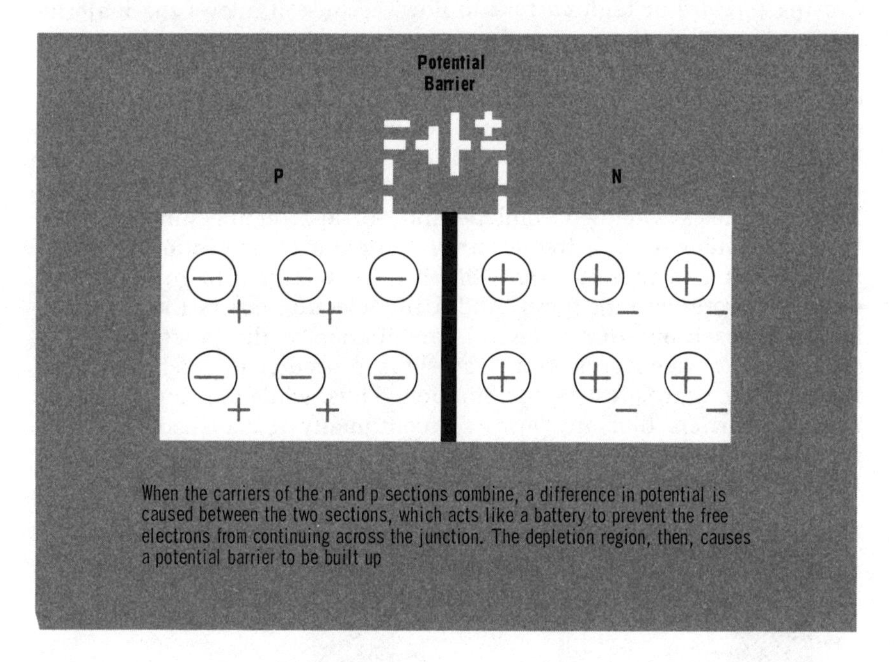

When the carriers of the n and p sections combine, a difference in potential is caused between the two sections, which acts like a battery to prevent the free electrons from continuing across the junction. The depletion region, then, causes a potential barrier to be built up

Because of the depletion region, the p and n sections of the diode now have equal and opposite charges. A voltage, a difference in potential, exists between the two sections. Since this difference in potential inhibits further electron-hole combinations at the junction, it is called a *potential barrier*. The difference in potential is the same as that between the terminals of a battery, and so the potential barrier is often shown as a small battery in the depletion region around the junction.

forward current flow

The natural tendency of the majority carriers, free electrons in the n section and holes in the p section, was to combine at the junction. This is how the depletion region and potential barriers were formed. Actually, the combination of electrons and holes at the junction allows electrons to move in the *same* direction in both the p and the n sections. In the n section, free electrons move toward the junction; in the p section, for the holes to move toward the junction, valence electrons move away from the junction. Therefore, electron flow in both sections is in the same direction. This, of course, would be the basis of current flow.

With the p-n diode alone, though, the action stops because there is no external circuit and because of the potential barrier that builds up. So, for current to flow, a battery can be connected to the diode to overcome the potential barrier. And the polarity of the battery should be such that the majority carriers in both sections are driven toward the junction. When the battery is connected in this way, it provides *forward bias*, causing *forward* or high *current* to flow, because it allows the majority carriers to provide the current flow.

Forward bias is obtained with the negative terminal of the battery connected to the n section, because the negative potential repels free electrons in the n section toward the junction. The positive terminal at the p section attracts valence electrons in the p section away from the junction, allowing the holes to move toward the junction. The free electrons and holes will then combine and be "lost" at the junction. But, for each combination, a free electron will enter the n section from the battery, and a valence electron will leave the p section to go to the battery. Therefore, current flows. And a free electron enters the n section to replace each one that is lost in combination at the junction. At the same time, a valence electron leaves the p section to produce a hole for each one that is lost in combination. Thus, while current flows, the majority carriers that are "lost" are continually replenished. The new free electrons and holes then drift toward the junction to continue the action. This is shown step-by-step on the following page.

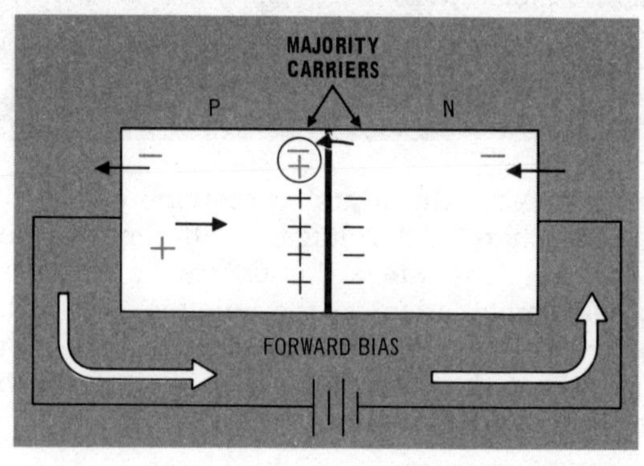

Forward or high current is produced in a p-n diode when forward bias is applied, which repels the majority carriers toward the junction to combine. This allows electrons to move in the same direction in both sections to produce current flow

forward current flow (cont.)

With forward bias applied to the p-n diode, the battery potential overcomes the potential barrier built up by the depletion region, and moves the majority carriers toward the junction.

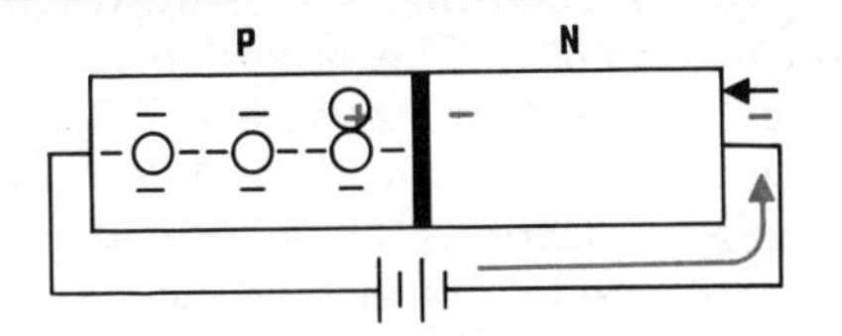

The hole on the p side of the junction attracts the free electron on the n side into it, and when they combine, both carriers are lost. At the same time, a free electron from the battery enters the n section to replace the one that recombined at the junction.

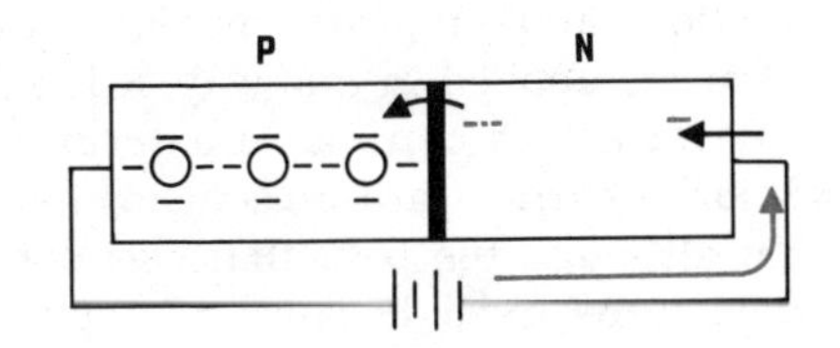

A free electron has an energy level close to or in the conduction band. When that free electron fills a hole at the junction, it becomes a valence electron. Its energy level, therefore, drops to the valence band. The energy it releases is transferred from atom to atom to the side connected to the positive battery. This raises the energy level of the valence electron to the conduction band, allowing the battery to attract it from the p section as a free electron. This produces a hole to replace the one that was lost at the junction.

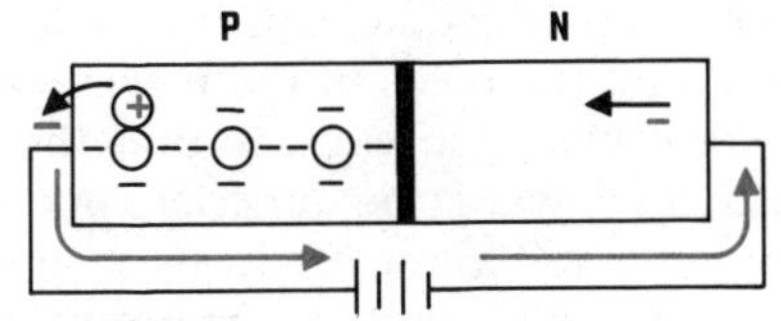

The positive attraction of the battery aids a valence electron from an adjacent atom in the p section to jump its bond and fill the new hole. The hole then moves to the adjacent atom toward the junction. The free electron that entered the n section is also repelled by the battery toward the junction.

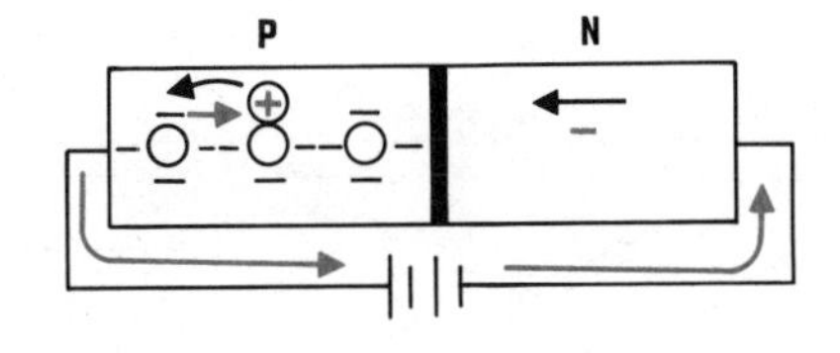

The hole and free electron continue to drift toward the junction until the original condition is recreated. Then, the free electron fills the hole, and the entire process is repeated.

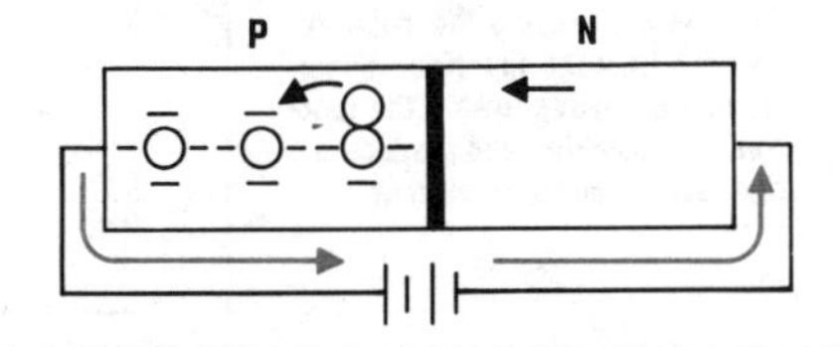

As shown by the step-by-step diagrams above, *forward current* flows as *free electrons* from the negative terminal of the battery, as *free electrons* through the n section, as *valence electrons* through the p section, and as *free electrons* to the positive terminal of the battery. It is usual practice, however, to refer to *hole flow* rather than *valence electron flow* in the p section.

reverse current flow

You learned that for forward current flow, the battery must be connected to drive the majority carriers toward the junction, where they combine to allow electrons to enter and leave the p-n diode. If the battery connections are reversed, the positive potential at the n side will draw the free electrons *away* from the junction. And the negative potential at the p side will attract the holes *away* from the junction. With this battery connection, then, the majority carriers *cannot* combine at the junction, and majority current cannot flow. For this reason, when a voltage is applied in this way, it is called *reverse bias*.

Reverse bias can cause a *reverse current* to flow, however, because *minority carriers* are present in the semiconductor sections. Remember, that although the p section was doped to have excess holes, some electrons were freed because of thermal agitation. Also, although the n section was doped to have excess free electrons, some electrons were freed to produce holes in the n section. The free electrons in the p section, and the holes in the n section are the minority carriers. Now, with reverse bias, you can see that the battery potentials repel the minority carriers toward the junction. As a result, these minority carriers can combine and allow electrons to enter and leave the p-n diode in exactly the same way that the majority carriers did with forward bias. However, since there are much fewer minority carriers than there are majority carriers, this minority current, or reverse current as it is usually called, is much less with the same voltage than majority, or forward, current would be.

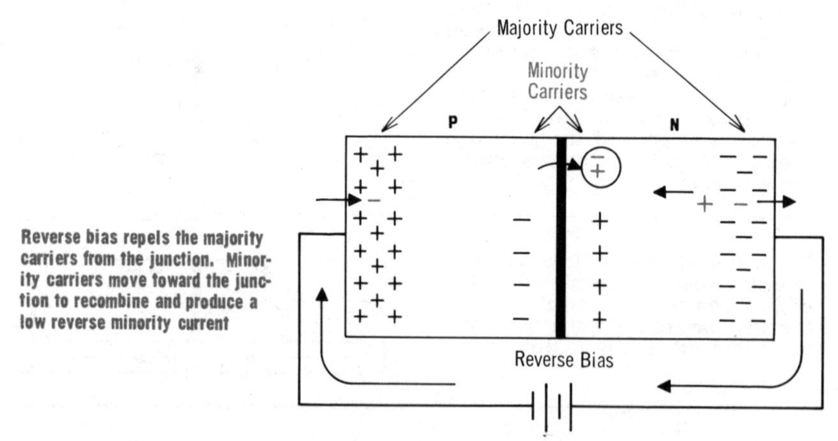

Reverse bias repels the majority carriers from the junction. Minority carriers move toward the junction to recombine and produce a low reverse minority current

The step-by-step description on the following page, describing minority or reverse current flow, will show how it is the same as majority or forward current. The interesting thing about the bias connections is that what is forward bias for the majority carriers is reverse bias for the minority carriers, and vice versa. This results because the majority and minority carriers in each section are opposites.

reverse current flow (cont.)

With reverse bias applied to the p-n diode, the battery potentials cause the majority carriers to drift away from the junction, so that they cannot take part in current flow. The minority carriers, however, are moved toward the junction.

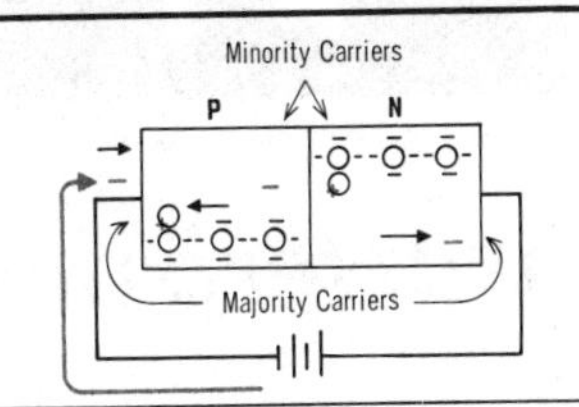

The hole on the n side of the junction attracts the free electron from the p side into it, and when they combine, both carriers are lost. At the same time, a free electron from the battery enters the p section to replace the one that recombined at the junction.

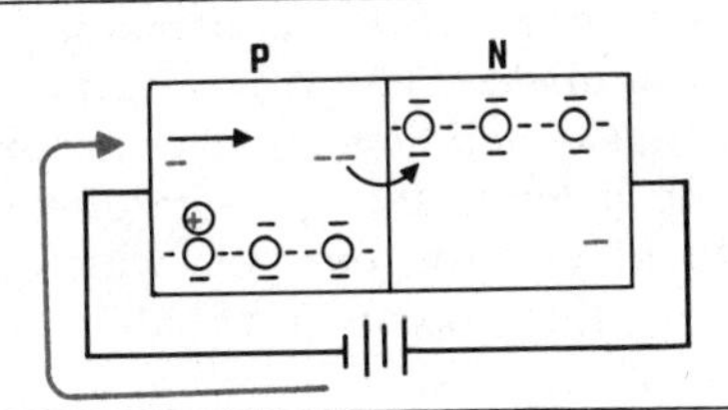

Since the free electron that crossed the junction had an energy level close to the conduction band, it had to release energy when it filled the hole and became a valence electron. The energy was transferred from atom to atom to the side connected to the positive battery terminal. This raised the energy level of a valence electron to the conduction band, allowing the battery to

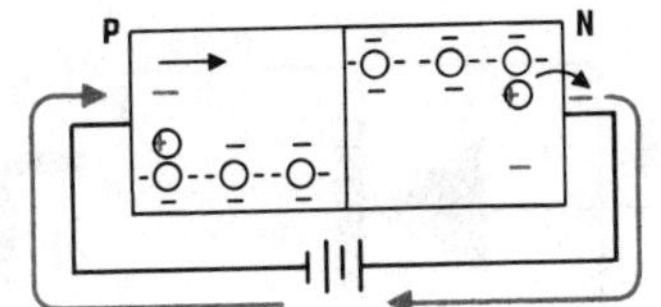

attract it from the n section as a free electron. This produces a hole to replace the one that was lost at the junction.

The positive attraction of the battery aids a valence electron from an adjacent atom in the n section to jump its bond and fill the new hole. The hole, then, moves to the adjacent atom toward the junction. The free electron that entered the p section is also repelled by the battery toward the junction.

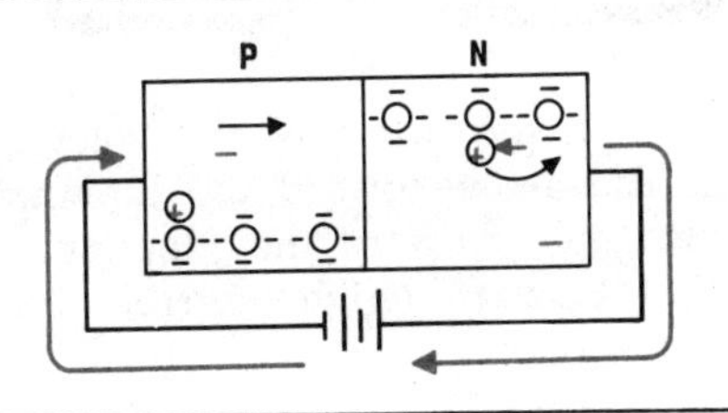

The hole and the free electron continue to drift toward the junction until the original condition is recreated. Then, the free electron fills the hole, and the entire process is repeated.

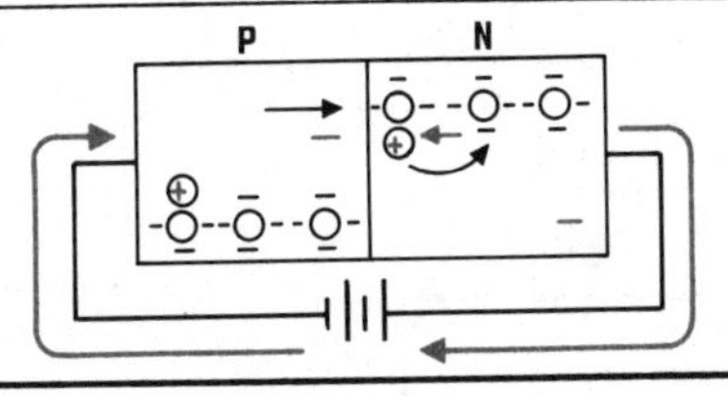

As shown by the step-by-step diagrams above, *reverse current* flows as *free electrons* from the negative battery terminal, as *free electrons* through the p section, as *valence electrons* through the n section, and as *free electrons* to the positive terminal of the battery. It is usual practice, however, to refer to *minority hole flow* in the n section rather than *valence electron flow*.

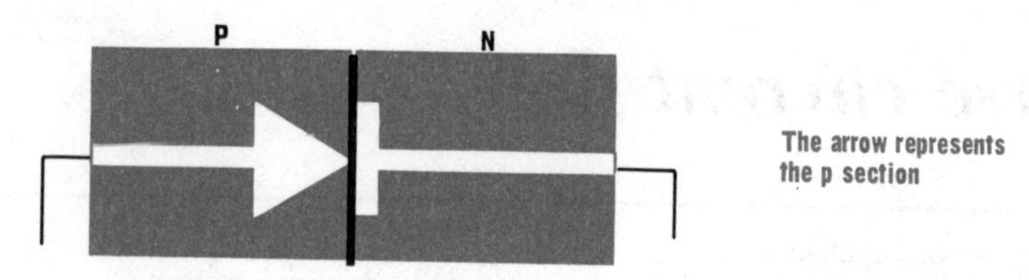

characteristics

The arrow in the semiconductor diode symbol represents the p section of the diode. The arrow points in the direction that the holes in the p section *should* flow for easy current. Therefore, since reverse bias attracts those holes in the opposite direction, the arrow will point in the direction of electron current, which in this case is minority current. The arrow, then, points in the direction of high resistance, or low (reverse) current.

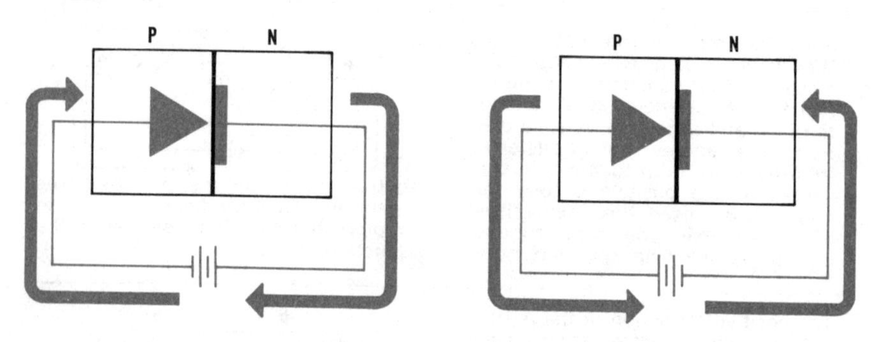

With forward bias, the holes in the p section move in the direction of the arrow toward the junction, and the electrons in the n section also move toward the junction. This allows majority current flow, so that forward (high) current and low resistance are opposite to the arrow.

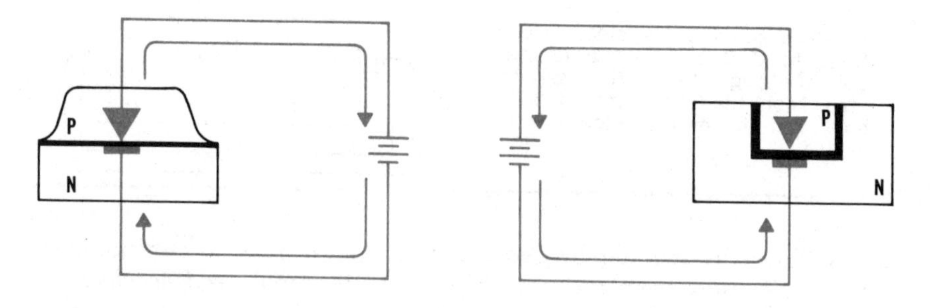

You can see, then, that to get forward current to flow, the negative side of the battery must be connected to the n side of the p-n diode; and vice versa.

the operating curve

Since the p-n diode's resistance changes according to the direction of current flow, it is a *nonlinear* device. Basically, its nonlinearity is dependent on the *polarity* of the applied voltage. For current in the forward direction, it has a resistance of only a few hundred ohms. In the reverse direction, its resistance is often close to 100,000 ohms.

The operating curve of the p-n diode shows how diode current changes with applied bias voltage, in both the forward and reverse directions. As shown, when reverse bias is applied, a slight reverse current flows, but this current increases only negligibly as the bias voltage is increased.

In the forward direction, though, considerably more current flows and the current for the most part, increases linearly as the bias voltage is increased. In the forward direction, then, the p-n diode can be considered a linear device over a large portion of its operating curve. The small part of the curve that is just above zero bias is nonlinear, as you can see. As explained earlier this results because both majority *and* minority current actually comprise the overall current. Since the minority carriers are low-energy carriers, majority current starts first, and then as the voltage is raised, minority current joins in, causing a nonlinear rise in current. But as the voltage is increased further, minority current becomes saturated since there are only few minority carriers. The curve then follows the majority current increase which is linear.

Because of the nonlinear *knee* of the curve, if a very small signal voltage is applied to the diode so that it only operates around the knee, the signal will be distorted. The signals must be large enough so that they operate mostly over the linear part of the curve.

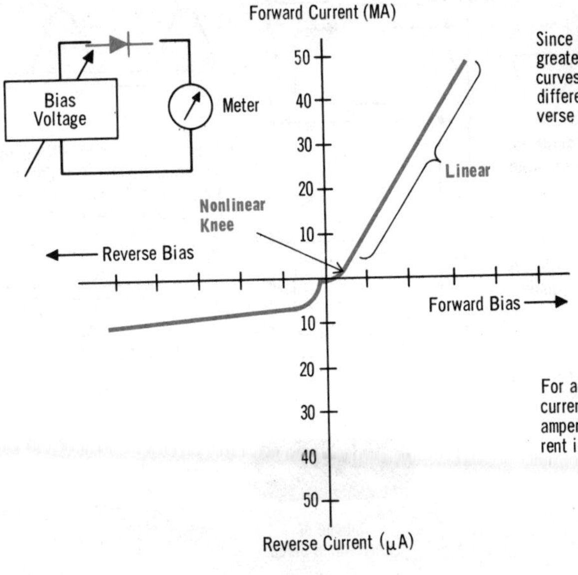

Since forward current is so much greater than reverse current, the curves are generally drawn to different scales to make the reverse current curve identifiable

For a small unit, the forward current curve indicates milliamperes, while the reverse current is shown in microamperes

the p-n diode rectifier

Since the p-n diode conducts current more readily in one direction than in the other, it can be used to convert an alternating current to a *unidirectional current.* When an a-c voltage is applied to a diode circuit, the diode will conduct relatively heavily when the polarity of the voltage produces a forward bias, but it will allow only a negligible current when the a-c polarity reverses to produce reverse bias. As a result, current flows essentially for only one-half cycle to produce a fluctuating dc at the output. This is similar to the way an electron-tube diode works. The electron tube, though, does not conduct at all in the reverse direction, whereas the p-n diode does, if only slightly. As a result, a very small portion of the blocked half of the a-c cycle *does* get through.

The half of the a-c sine wave that causes forward bias depends on the way the diode is connected. By reversing the diode connections, either the positive or the negative half cycle can be passed. Selenium, copper oxide, silicon, and germanium rectifier diodes work in this way. Quite often, rectifier diodes that are used in electronic power supplies have a plus (+) sign marked on one side of the rectifier. The plus sign merely indicates that the fluctuating d-c voltage at that side will be positive when the a-c voltage is applied to the other side. Most small signal diode rectifiers do not have this marking. Instead, they might contain the diode symbol to show the high-resistance direction. With a p-n diode, the *n* section is often called the *cathode,* and the *p* section is commonly called the *anode* for ease of comparison to the operation of tube diodes.

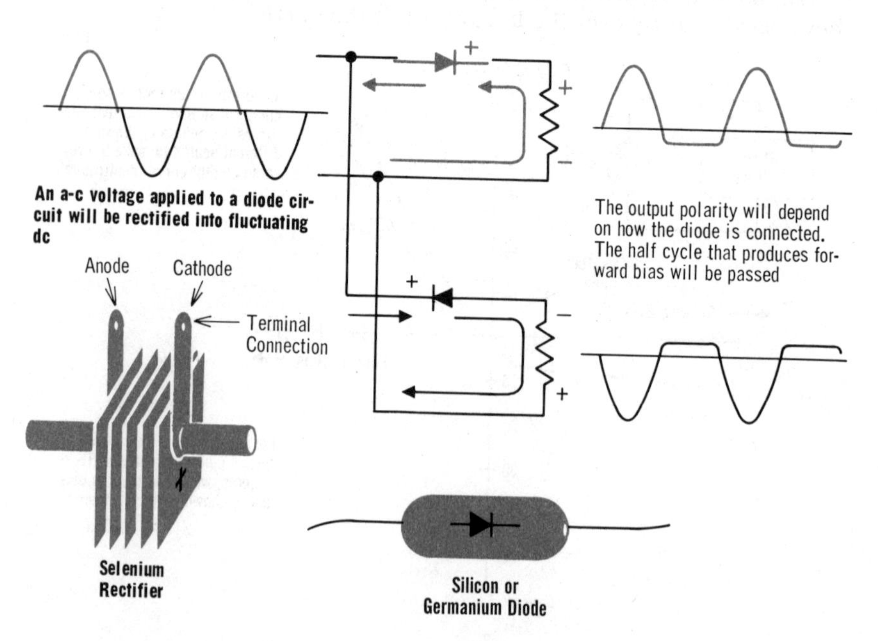

An a-c voltage applied to a diode circuit will be rectified into fluctuating dc

The output polarity will depend on how the diode is connected. The half cycle that produces forward bias will be passed

Selenium Rectifier

Silicon or Germanium Diode

junction capacitance

A p-n diode can be thought of as similar to a capacitor, since the semiconductor segments are similar to electrodes separated by a dielectric (the junction). With the diode biased in the forward direction, the junction is a low-resistance path, and so is inefficient as a dielectric, and the *junction capacitance* is nil. When the diode is biased in the reverse direction, however, the junction resistance becomes significant and functions more efficiently as a dielectric. An actual junction capacitance exists, and this capacitance can be used in a circuit in the same way as the capacitance of a regular capacitor.

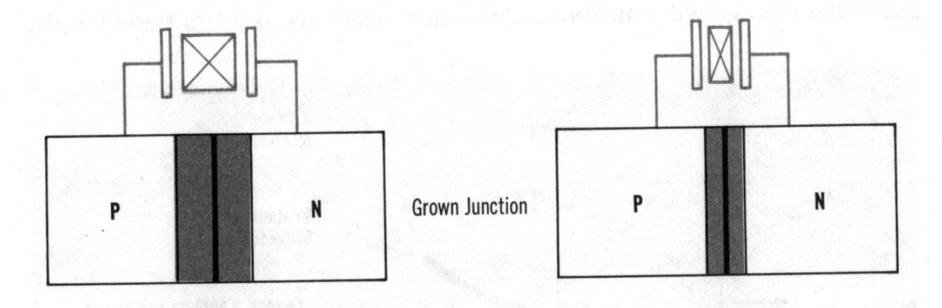

A p-n diode biased in the reverse direction acts like a capacitor, with the junction and depletion region acting as the dielectric. At low reverse voltages, the depletion region is narrow and the junction capacitance is high. Varactor diodes are specially made with long junctions and narrow depletion regions for accurate control of the junction capacitance

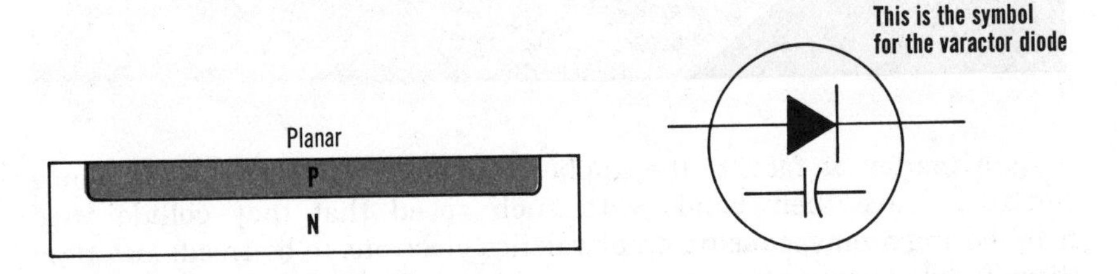

This is the symbol for the varactor diode

In a p-n diode, the depletion region, which is essentially void of majority carriers, forms part of the dielectric with the junction. Therefore, since the reverse voltage repels majority carriers away from the junction, higher voltages will widen the depletion region, thereby effectively moving the electrodes farther apart, so that junction capacitance decreases as reverse bias is increased. By adjusting the reverse bias, certain specific values of junction capacitance can be controlled, so that the diode can be used as a capacitor. Certain p-n diodes are deliberately manufactured to take advantage of this characteristic. They are designed to produce specific capacitances, and even a range of *variable capacitance*, for use in tuned circuits. Such p-n diodes are known as *varactor diodes*, *varicaps*, or *voltacaps*. They are operated in the *reverse direction* only.

avalanche breakdown

Remember, minority current also forms a small part of the forward current that flows through a diode; but, when the diode is operated past the knee of the curve, the minority current *saturates* because there are only a few minority carriers compared to the number of majority carriers. Minority carriers in the n section are holes that result from valence electrons being freed. In the p section, these freed valence electrons are the minority carriers. Now, during normal operation, the number of minority carriers is stable. But, if you recall, the valence electrons can be set free if enough energy is applied. This can happen if too high a forward bias is applied to a p-n diode. The high attraction force of the excessive voltage will cause many valence electrons to jump their bonds.

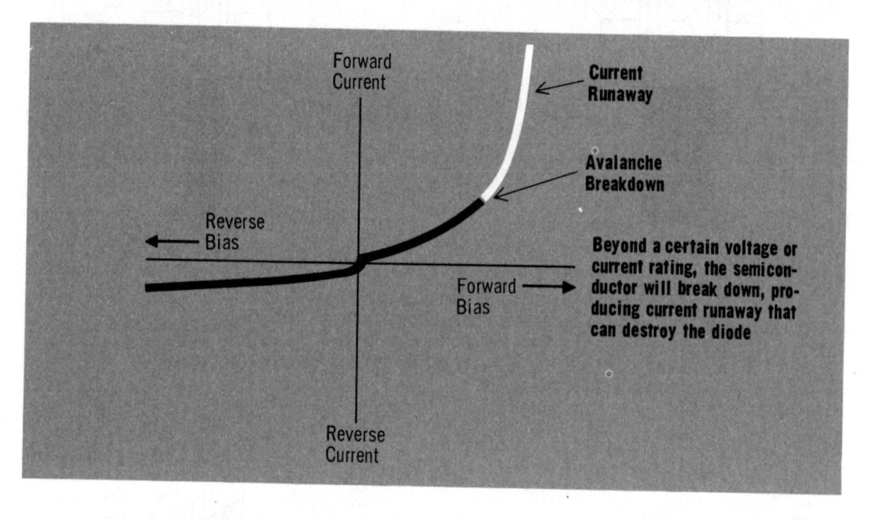

As a matter of fact, if the applied voltage is sufficient, the valence electrons leave their bonds with such speed that they collide with neighboring atoms, causing other valence electrons to be freed, and then these freed electrons can continue the action. The action is regenerative, and an *avalanche* of new carriers is produced. In this way, the number of minority carriers can be increased to where they actually exceed the majority carriers. The forward current then can suddenly rise, and if the diode is not designed to take the current surge, the diode can be destroyed. The point at which this occurs is called the *avalanche breakdown* voltage, or *runaway* voltage. Normal rectifier and signal diodes are rated with applied voltage limits to prevent this from happening.

Minority current runaway can also occur if the temperature of the p-n diode becomes too high because of *thermal agitation*. This produces *thermal runaway*. Thermal runaway will also take place if too much forward current flows through the diode, since excessive current will heat up the diode. Therefore, diodes also have a current limit rating.

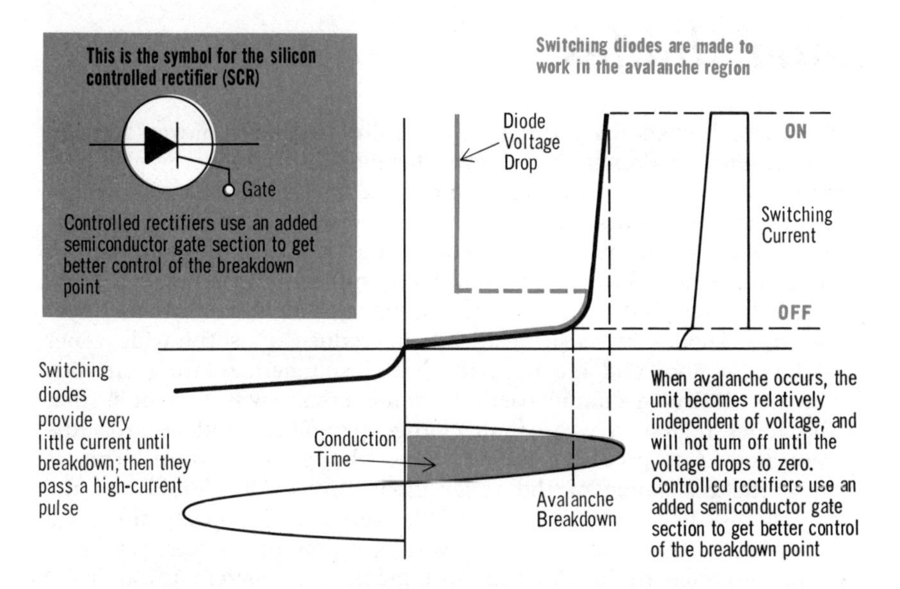

switching diodes and controlled rectifiers

Normally, rectifier and signal diodes are destroyed when avalanche breakdown or thermal runaway occurs. Some diodes, however, are manufactured with large semiconductor segments to survive the sudden current surge. This almost instantaneous current change is desirable in circuits that need to be switched on and off electronically, rather than by switches or relays. *Electronic switches* are much faster and longer lasting than mechanical or electromechanical switches.

Diodes designed to work in the *avalanche region* are called *switching diodes*. Such diodes are made with very few majority carriers, and so allow very little forward current until avalanche breakdown occurs. Then, they allow a very high current to pass, energizing a circuit. These diodes are not usually designed to rectify or pass a given waveform. They merely act as high-current switches.

At the avalanche breakdown point, the resistance of the diode drops sharply, so that the voltage dropped by the diode in a circuit also drops to a low value. However, once avalanche current starts, the resistance of the diode remains low, and becomes relatively independent of the voltage across the p-n unit. The switching diode then works similarly to the gas tube diode you learned about in Volume 3. The diode voltage must drop to just about zero before it will stop conducting.

Special controlled rectifier diodes have an additional semiconductor element, called a *gate*, added to give better control of the avalanche breakdown point. This unit, called an SCR, works like a thyratron tube, and is described later.

zener diodes

Avalanche breakdown is the term generally used for runaway current. However, since avalanche breakdown depends on high-energy electrons colliding with atoms, it is usually preceded by another breakdown in conduction. This occurs when the applied voltage is sufficient to cause valence electrons to jump their bonds and increase the number of carriers, but when the voltage is still not sufficient to allow the high-energy collisions that bring about avalanche. This is known as *zener breakdown*. Diodes are also designed to produce a useful wide zener breakdown region, and are used in special voltage-regulating circuits.

The one desirable characteristic of zener breakdown current is that in the zener region, a very small change in voltage still controls the current flow, but produces a very large change in current. This is because the resistance of the zener diode drops considerably as the voltage across it is increased beyond the zener breakdown point. As a result, when a zener diode is used with a dropping resistor, wherever the voltage across the diode tends to increase, the current through the diode rises out of proportion and causes a sufficient increase in voltage drop across the dropping resistor to lower the output voltage back to normal. Similarly, when the voltage across the diode tends to decrease, the current through the diode goes down out of proportion so that the dropping resistor drops much less voltage to raise the output voltage to normal. The zener diode is always used in the *reverse* direction.

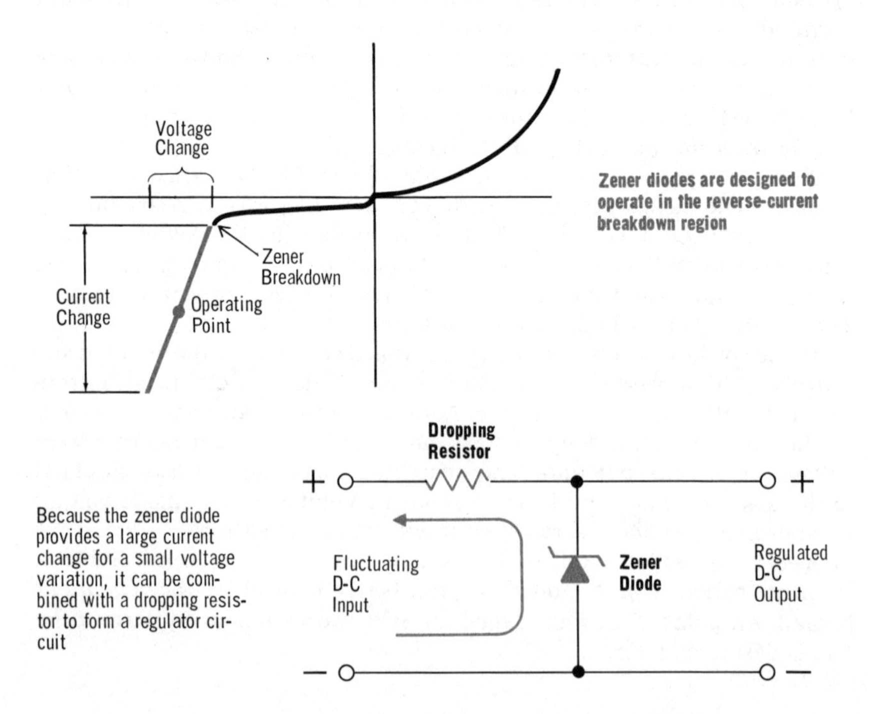

Zener diodes are designed to operate in the reverse-current breakdown region

Because the zener diode provides a large current change for a small voltage variation, it can be combined with a dropping resistor to form a regulator circuit

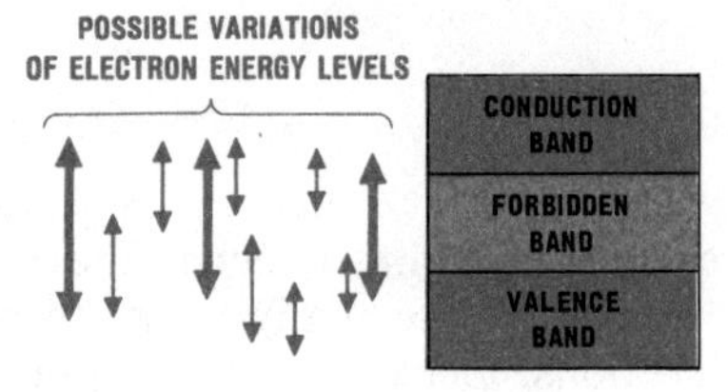

Some electrons can gain enough energy to go from the valence energy level to the conduction level; and in doing so can go from the p to the n side. This is the tunneling effect

the tunnel effect

Zener and avalanche breakdown depend on valence electrons receiving sufficient energy to break from their covalent bonds and increase the number of available current carriers, particularly minority carriers. For zener or avalanche operation, this is easily accomplished with a high applied voltage. However, some covalent electrons go from the valence energy band to the conduction band with little or no applied voltage.

Seemingly, an electron whose energy level is in the valence band cannot travel out of a covalent bond until its energy level is raised to the conduction band. However, although energy level diagrams are shown with clearly defined lines separating the valence, forbidden, and conduction bands, the energy levels of individual electrons cannot be that clearly defined.

Actually, a valence electron can have an energy level anywhere within each band. In addition, since there is electron activity within the semiconductor, because valence or free electrons drift, filling or leaving holes, the energy level of individual electrons can continually shift from one band to the next. The reason for this energy level shift is that when a free electron fills a hole to become a valence electron, it releases energy to get from the conduction to the valence band. Generally, this energy is passed to another valence electron that can leave its bond as its energy level goes from the valence to conduction band.

Many valence electrons have energy levels in the forbidden band. Thus, they are not free electrons, but rather valence electrons that can travel freely from hole to hole. Ordinarily, not many drift at random this way, but when they do, they usually release energy to another valence electron to put its energy level in the forbidden band so that it can jump into another hole. However, when a p-n semiconductor is *heavily doped,* and has many majority carriers and ions, the hole and valence electron random drift is heavy. As a result, it is not uncommon for a large number of electrons to fill holes and release energy to only a few other valence electrons. These few valence electrons, then, have their energy levels raised considerably so that they can go into the conduction band, and even cross from the p to the n section as minority carriers to fill a hole, even with little or no applied voltage. This action, which seems to allow a valence electron to go from the valence band to conduction band and cross a potential barrier without enough applied external energy is called the *tunnel effect* because it seems as though the valence electron "tunnels" through the forbidden band.

tunnel diodes

A diode that makes use of the tunneling effect is the *tunnel diode.* It is very heavily doped so that there are many majority carriers and ions in the semiconductor sections. Because of the large number of carriers, most are not used during the initial recombination that produces the depletion region. As a result, the depletion region is very narrow, producing a thin junction that is easily crossed by electrons.

Because of the large number of carriers, there is much drift activity in the p and n sections, causing many valence electrons to have their energy levels raised closer to the conduction region. Therefore, it takes only a small applied forward voltage to cause conduction. As forward bias is *first increased,* diode current rises rapidly; after many carriers start participating in current flow, the random activity of the free electrons filling holes is reduced considerably, so there is much less tendency for valence electrons to be raised in energy to the conduction band. Therefore, the tunnel effect is reduced, and majority carriers make up most of the current flow.

The sharp reduction of the available minority carriers reduces the overall current flow, so that current flow starts to *decrease* as diode applied voltage is increased. As the voltage is further increased, the tunneling effect plays less and less of a part until a *valley* is reached when the p-n unit starts to act as a normal semiconductor diode; the current then rises with voltage. The area after the *tunneling current* reaches its peak is called the negative resistance region because current goes down as voltage is raised. Because of the tunneling effect, the tunnel diode has a quick response, and can be used as a good electronic switch between the peak and valley of current. Also, the negative resistance region allows the diode to be used as an oscillator.

By its nature, the tunnel diode has a rather high reverse current, but operation under this condition is not generally used.

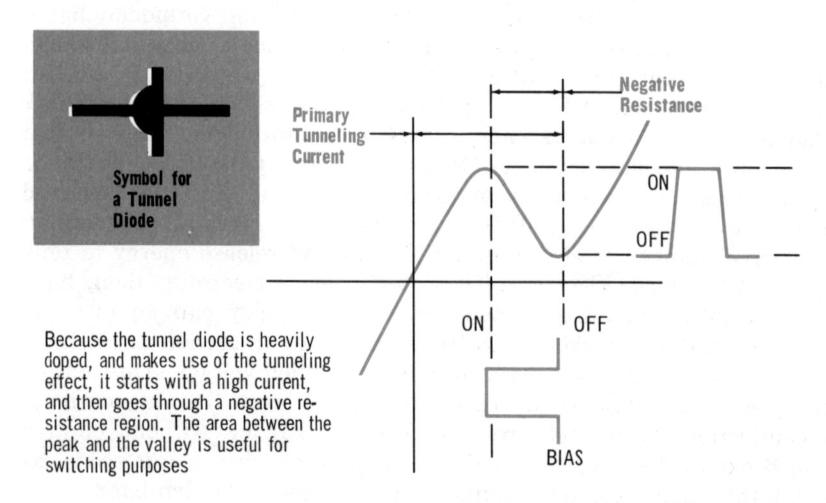

Because the tunnel diode is heavily doped, and makes use of the tunneling effect, it starts with a high current, and then goes through a negative resistance region. The area between the peak and the valley is useful for switching purposes

summary

□ In a p-n diode, electron-hole combinations cause a depletion area at the junction. □ A potential barrier is produced across the junction, which inhibits current flow through the junction. □ Forward bias overcomes the potential barrier to allow majority carriers to flow. □ Reverse bias aids the potential barrier by repelling majority carriers away from the junction, and drives minority carriers toward the junction to produce a low reverse current.

□ The arrow in a p-n diode symbol represents the p, or anode section, and points in the direction of high resistance. Current flows easily only when it is opposite to the arrow. □ The p-n diode rectifies ac by passing only half of the cycle.

□ The p-n junction acts as a dielectric, and so capacitance appears across the junction. This capacitance can be made to vary with voltage to produce a variable capacitance diode (varactor).

□ Too high a voltage in the forward direction will cause avalanche breakdown. □ Too high a reverse voltage will cause zener breakdown. □ Switching diodes are designed to work in the avalanche region, and voltage-regulating, or zener diodes are designed to operate in the zener region.

□ The tunnel effect is the movement of valence electrons from the valence energy band to the conduction band with little or no applied energy. □ The tunnel effect causes a negative resistance region, where an increasing voltage results in a decreasing current, and vice versa. Tunnel diodes are used in switching and oscillator circuits.

review questions

1. How is a *grown junction* diode made?
2. The *potential barrier* increases with ________ bias and decreases with ________ bias.
3. Forward bias causes ________ carriers to carry the current flow.
4. *Forward* current is in the direction of the *arrow* in a diode symbol. True or false?
5. Does current flow during both cycles when a diode *rectifies* a sine wave? Why?
6. Functionally, the p-n diode acts like what *reactive* component?
7. Define *avalanche breakdown. Zener breakdown.*
8. Normally, voltage regulating diodes work in the ________ region.
9. Describe the tunnel effect, and how it causes negative resistance.
10. Draw the symbols for: (1) a p-n diode, (2) a varactor diode, (3) a switching diode, (4) a zener diode, and (5) a tunnel diode.

the transistor

Until now, we have been mainly concerned with semiconductor materials and semiconductor diodes. These materials and devices have unique properties that make them useful in many applications. However, they do have limitations that prevent them from being used in many other applications.

The biggest limitation of the semiconductor diode is that most cannot amplify a signal in a practical manner. A solid semiconductor is applied mostly in a way that makes use of its nonlinear resistance characteristics, and a p-n diode is used for rectification as well. When amplification is needed, another type of semiconductor device is used: the *transistor*. There are various types of transistors. The most common types are called the junction and field-effect transistors. The junction transistor will be discussed first.

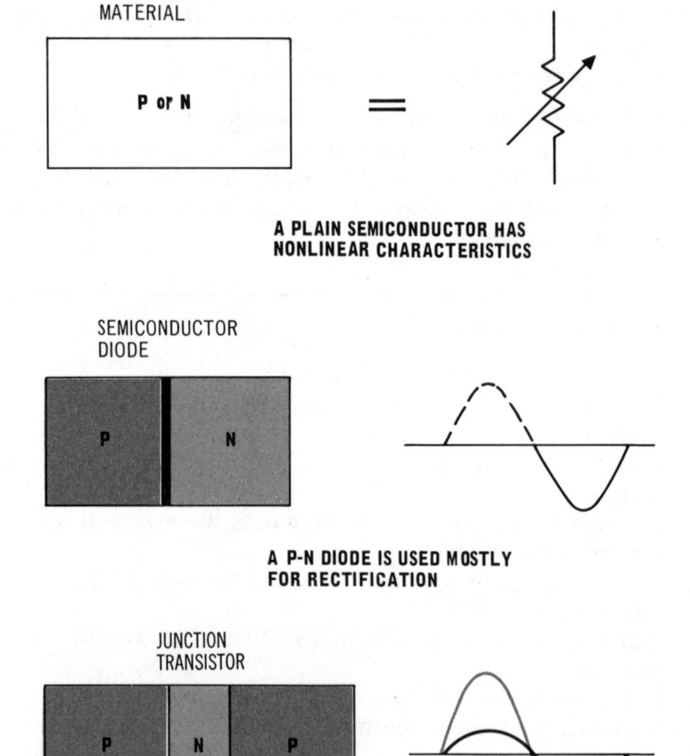

A PLAIN SEMICONDUCTOR HAS NONLINEAR CHARACTERISTICS

A P-N DIODE IS USED MOSTLY FOR RECTIFICATION

THE TRANSISTOR, HOWEVER, CAN BE USED FOR SIGNAL AMPLIFICATION

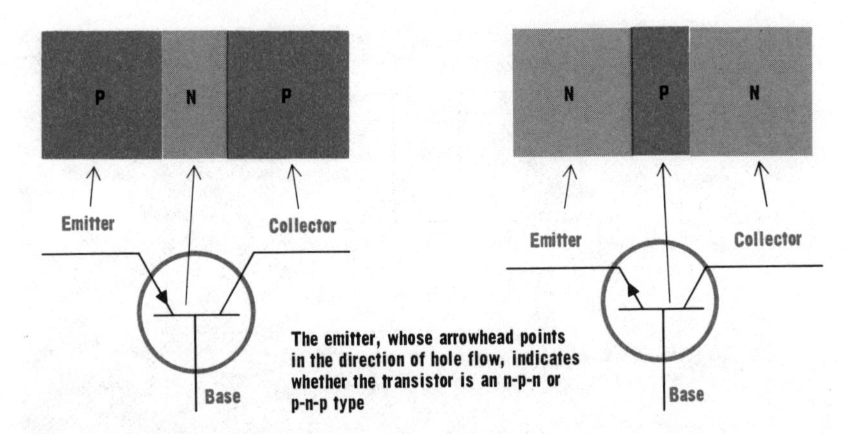

The emitter, whose arrowhead points in the direction of hole flow, indicates whether the transistor is an n-p-n or p-n-p type

the basic junction transistor

The basic junction transistor is produced when another semiconductor element is added to the simple p-n diode. The transistor, then, is a *three-element semiconductor*. The three elements are combined so that the two outer elements are doped with the same type of majority carriers, while the element that separates them has the opposite majority carrier. A transistor, then, can be an n-p-n or a p-n-p type.

The three elements of the transistor are the *emitter*, the *base*, and the *collector*. The *emitter* supplies the majority carriers for transistor current flow, and the *collector* collects the current for circuit operation. The *base* provides the junctions for proper interaction between the emitter and collector.

The emitter is shown schematically by an arrow that points in the direction of *hole* flow. Since the emitter is said to inject majority carriers into the base, a *p*-type emitter is shown with the arrow pointing *to* the base; an *n*-type emitter has the arrow pointing *away* from the base to show that electrons are being injected.

The manufacture of transistors is similar to that of diodes. The illustration at the top of the page shows the *grown junction* type of transistor, and those below show the mesa and planar types. These are all junction transistors and are also called *bipolar transistors*. This designation is used to differentiate them from field-effect transistors, which are called *unipolar transistors*. This will be explained later.

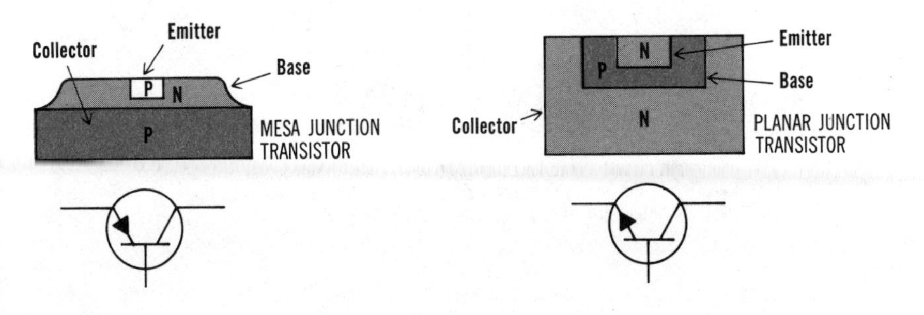

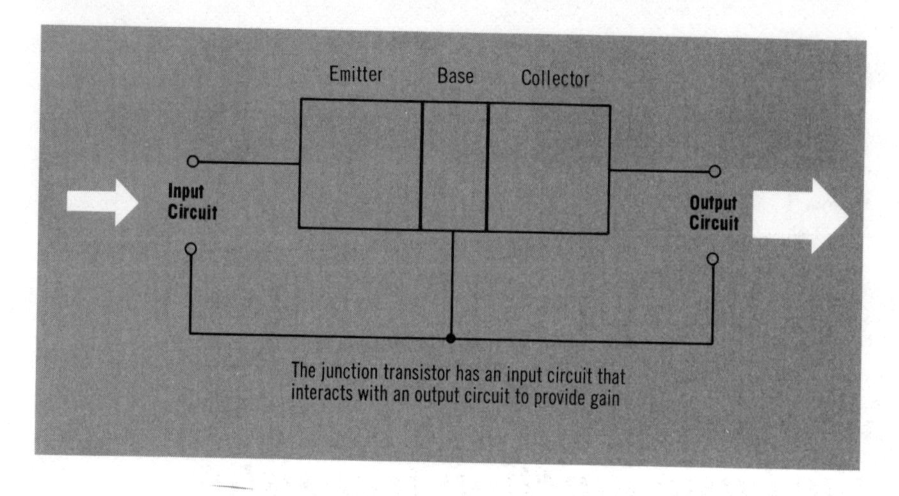

The junction transistor has an input circuit that interacts with an output circuit to provide gain

input and output circuit interaction

Basically, the transistor amplifies in a manner similar to that of an electron tube. It is set up with two circuits: an *input circuit*, and an *output circuit*. The two circuits are arranged so that they interact. The current that flows in the input circuit controls, to a large extent, the current that flows in the output circuit. Therefore, when a signal is applied to the input circuit, it produces a corresponding input current flow. This, in turn, determines the current that flows in the output circuit. Because of this, if the input circuit is made a low-voltage circuit, and the output circuit is made a high-voltage circuit, then small signal voltage inputs can produce higher signal voltage outputs.

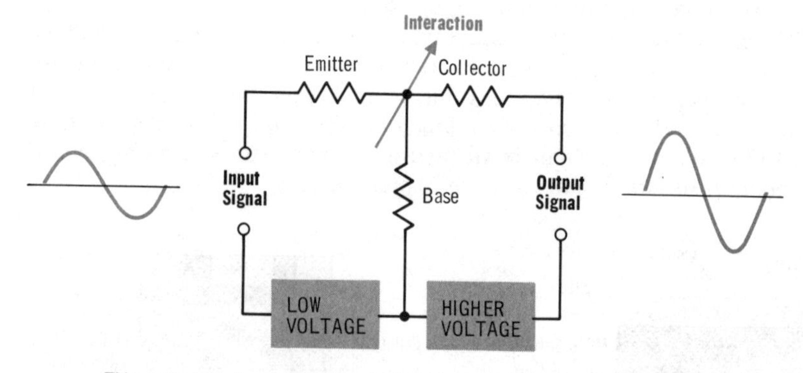

This equivalent circuit shows how the input interacts with the output. The input signal applied to the low-voltage input circuit controls the resistance of the higher-voltage output circuit to produce an amplified signal

junction transistor biasing

One reason that the input and output circuits of a transistor interact is because the base of the transistor is common to both circuits. Another factor that controls interaction is the types of bias used in both circuits.

The input circuit is provided by the emitter and base segments, and the output circuit is provided by the base and collector segments. Since the input circuit current must be determined by the input signal voltage, the emitter and base must be biased in the *forward* direction so that emitter-base current will follow the signal voltage. However, since the output circuit current must be controlled by interaction with the input circuit, the output circuit current should be relatively independent of circuit voltages. This is accomplished by biasing the base-collector circuit in the *reverse* direction. Because reverse bias produces very little current flow, the output current is hardly affected by it.

The input, or emitter-base circuit, is forward biased; and the output, or collector-base circuit, is reverse biased

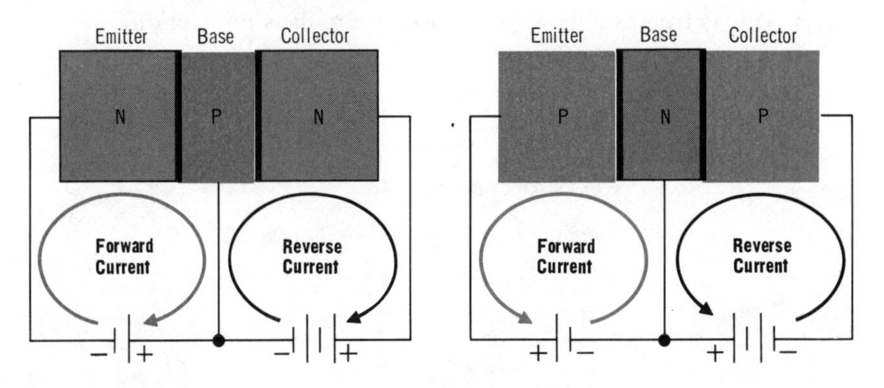

The forward emitter-base bias is generally lower than the reverse collector-base bias, and both currents are almost equal. Also, since both currents flow through the base in opposite directions, they almost cancel and there is very little resultant base current

Ordinarily, the collector current should be very low compared to the emitter current, but it is made high in two ways. The first is by use of a relatively high collector voltage. The second, and most important, way is that, as shown on the diagram, the bias polarities of both circuits are such that they aid emitter current to flow through the collector as well. As a result, collector current is usually almost equal to emitter current. This is how the output current is controlled by the input current.

You might have noticed that the emitter and collector electron currents appear to be going in the same direction. They do in the circuit, but they go in opposite directions through the junctions. Notice that in the n-p-n type, the emitter-base electron current crosses the junction from n to p, but the collector-base electron current crosses from p to n.

n-p-n input circuit operation

Let's see how the n-p-n transistor works with only the emitter-base input circuit connected. You should recall that in p-n diodes *forward* current across a junction is carried out by *majority* carriers, recombining at the junction to allow electrons to flow in and out of the semiconductor. In an n-p-n transistor, the same is true for the input circuit, since the emitter-base junction is forward biased. The bias polarity repels the free electrons in the n-type emitter toward the junction, where they cross to fill holes in the p-type base. However, the transistor differs from the diode in that the two segments on opposite sides of the junction are not equally doped nor are they the same size.

The base is thinner and is doped much less than the emitter. Therefore, it has much less majority carriers than the emitter. As a result, there are more free electron majority carriers that go from the emitter to the base than there are hole majority carriers in the base. So, most of those free electrons cannot recombine; they just accumulate in the base, and restrict the electron current flow in the circuit to a very small value. This is true as long as the collector circuit is not working.

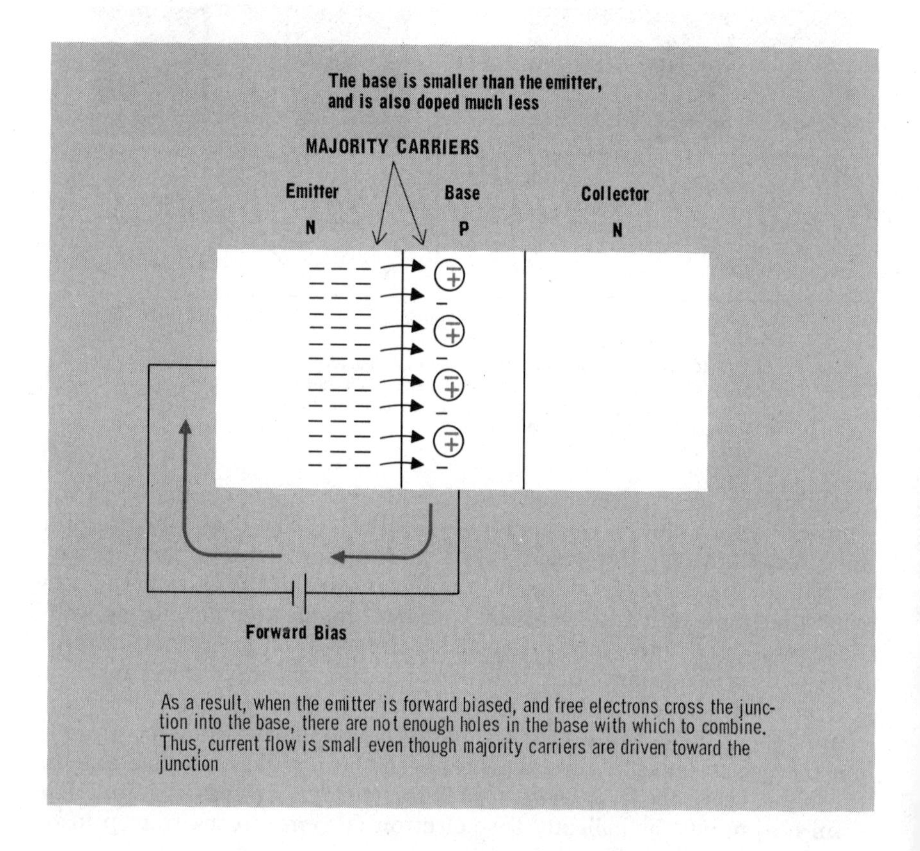

As a result, when the emitter is forward biased, and free electrons cross the junction into the base, there are not enough holes in the base with which to combine. Thus, current flow is small even though majority carriers are driven toward the junction

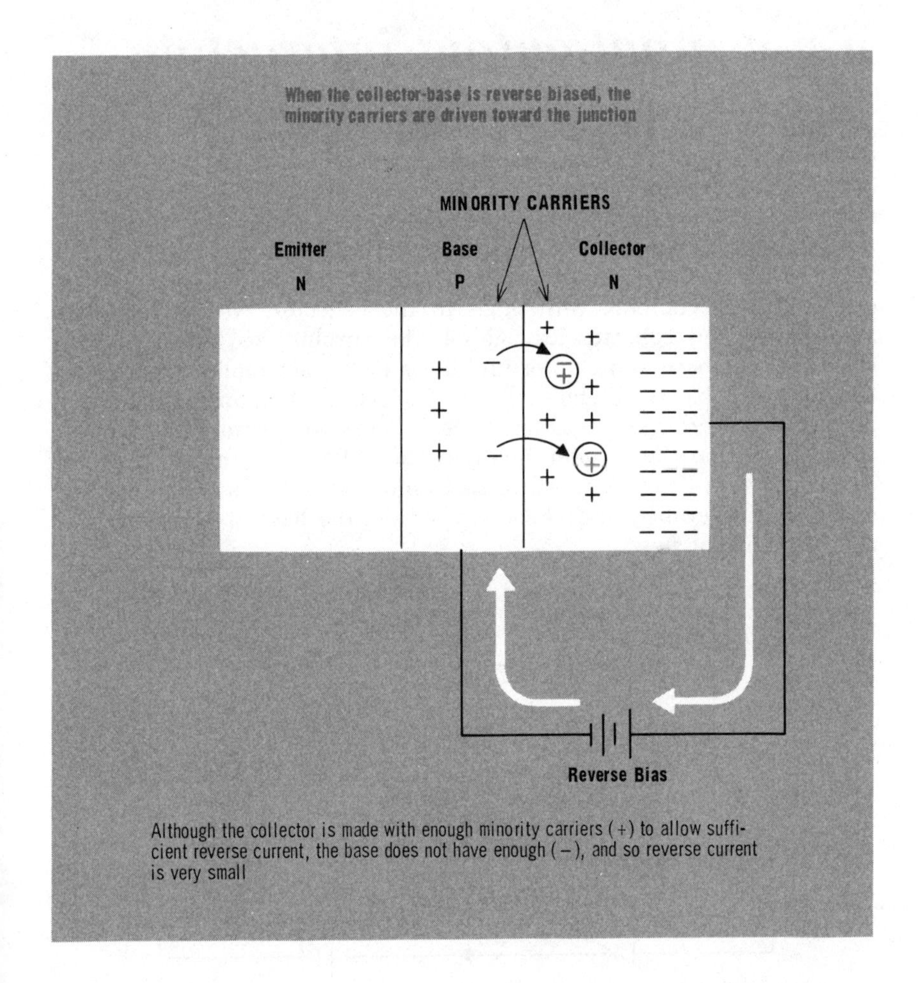

Although the collector is made with enough minority carriers (+) to allow sufficient reverse current, the base does not have enough (−), and so reverse current is very small

n-p-n output circuit operation

As before, let us examine this operation with only the collector-base circuit connected. As with the p-n diode, *reverse* current in a transistor is carried out with *minority* carriers. In the n-type collector, minority carriers are holes; in the p-type base, minority carriers are free electrons. Therefore, the reverse current will depend on the number of free electrons that cross from the base to recombine with holes in the collector.

The collector is made so that it contains a sufficient number of minority carriers to provide a usable reverse current. But, as with emitter-base current, the base also has only a few minority carriers. Therefore, there is very little recombining of minority carriers at the collector-base junction, and so very little reverse current flows. This is true only as long as the emitter circuit is inoperative.

n-p-n input-output interaction

When both the emitter and collector are connected, the operation of the n-p-n transistor changes considerably. Remember that emitter-base forward current was kept low because the base did not have enough majority carriers (holes) to recombine with the free electrons from the emitter. In addition, collector-base reverse current was kept low because the base also did not have enough minority carriers (free electrons) to recombine with holes in the collector. Now, when both the emitter and collector are biased, the surplus free electrons that cross into the base and cannot find holes to fill, accumulate there, and become available to fill holes in the collector. Therefore, since more majority carriers (free electrons) can leave the emitter to *diffuse* through the base and enter holes in the collector, a higher forward emitter current flows. Also, since more minority carriers (holes) in the collector can be filled by free electrons from the base, a higher reverse collector current flows.

When both emitter and collector are biased, the surplus free electrons from the emitter can pass through the base to combine with holes in the collector

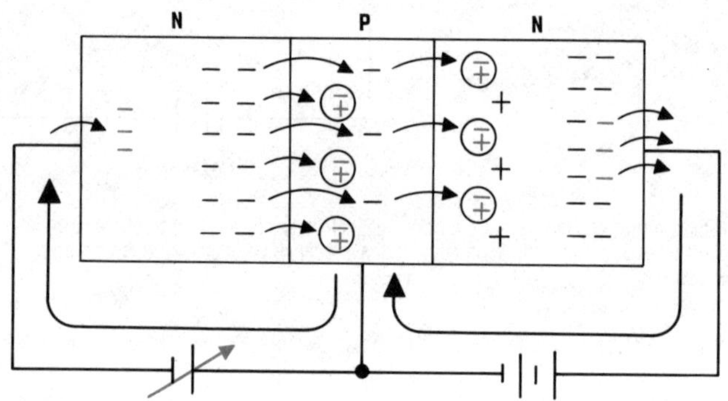

The amount of collector current that flows depends on the number of free electrons supplied by the emitter. If the emitter bias voltage is raised or lowered, emitter current will change, causing collector current to change, too

Since the collector-base is reverse biased, the amount of collector current that flows depends only slightly on the collector voltage. It is determined more by the number of surplus free electrons that the emitter supplies to the base. The emitter-base, on the other hand, is forward biased, and so its current flow will go up or down as emitter bias is raised or lowered. Accordingly, this will also allow the emitter to supply more or less free electrons to the base to cause the collector current to rise and fall. In this way, any change in emitter current will cause a corresponding change in collector current.

It is interesting to note that although the free electrons are majority carriers in the emitter (n), they become minority carriers in the base (p). Because the junction transistor uses two kinds of carriers (majority and minority), it is called *bipolar*.

n-p-n gain

To make the n-p-n current characteristics useful, it is connected in a way that allows a signal input voltage to be applied to the emitter, and a signal output voltage to be taken from the collector. The input signal voltage, then, will either aid or oppose the forward bias of the emitter circuit. When it aids the bias, emitter current goes up, and vice versa. And, when the signal voltage causes emitter current to change, the collector current changes, too. If a load resistor is put in the collector circuit, any change in collector current will produce a changing voltage drop across the load resistor.

Since the emitter circuit is forward biased, it is a low-resistance circuit. The collector circuit, though is reverse biased, and so is a high-resistance circuit. The load resistor, then, can be made a high value. As a result, the voltage changes across the load resistor will be greater than the input signal voltage variations. You can see this more easily when you realize that the collector current is about equal to the emitter current. This is because the base has so few carriers that only about 2 to 5 percent of the emitter current flows down the base; about 95 to 98 percent of the emitter carriers form the collector current. So, the same current passing through two different resistances will naturally cause a higher voltage drop across the larger resistance.

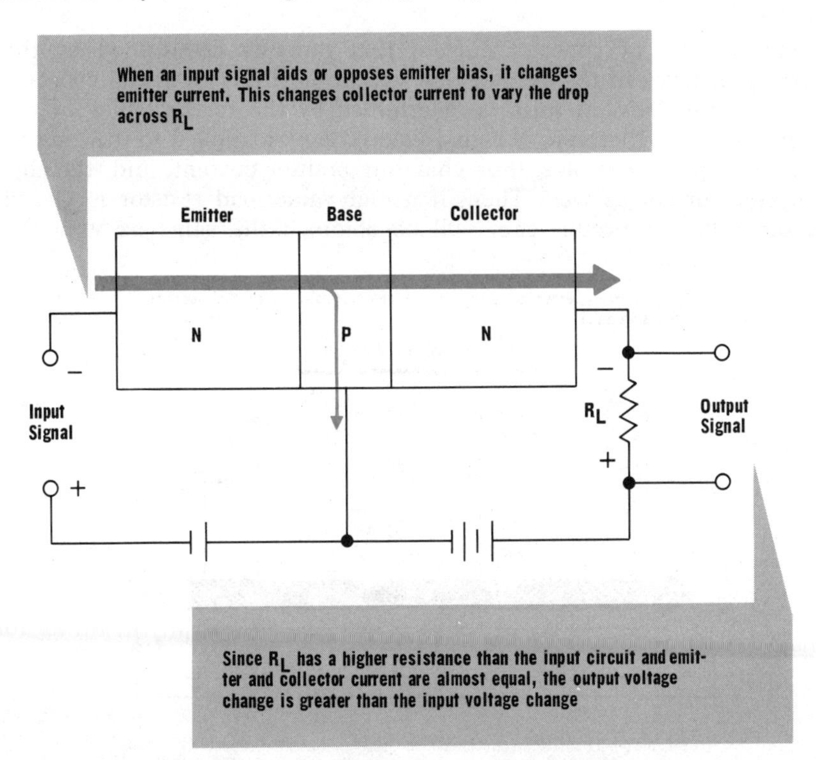

p-n-p operation

The p-n-p transistor operates similarly to the n-p-n type, except that the majority and minority carriers in the p-n-p transistors are opposite. The emitter-base segments are forward biased and the collector-base segments are reverse biased. The forward bias on the emitter attracts valence electrons away from the junction, causing holes, the majority carriers, to drift toward the junction. The free electron majority carriers in the base then cross the junction to fill holes and allow circuit electron flow to enter the base and leave the emitter. However, the base is made thin and is only slightly doped, so that it does not ordinarily have enough majority carriers to fill holes in the emitter; so emitter current tends to be low. The excess holes in the emitter accumulate at the junction.

Since the collector is reverse biased, the voltage polarity drives the free electrons, which are minority carriers, toward the junction. These free electrons cross the junction to fill minority carrier holes in the base, thus allowing free electron circuit current to flow. However, the base also has very few holes, so that most of the free electrons from the collector tend to accumulate there and restrict current flow. But, these free electrons now become majority carriers in the base, and are attracted across the junction to fill the accumulated holes in the emitter. Because of this, both a high emitter and collector current are allowed to flow.

The amount of collector current that flows is determined by the number of holes in the emitter that accept electrons from the collector via the base. This, in turn, is determined by the forward bias in the emitter circuit. Therefore, a signal voltage can be applied to the emitter to aid or oppose the bias, thus changing emitter current, and, thereby, collector current as well. Then, if a high-value load resistor is placed in the collector circuit, gain will be obtained the same as with the n-p-n transistor.

The p-n-p transistor works the same as the n-p-n transistor, except that opposite carriers are used

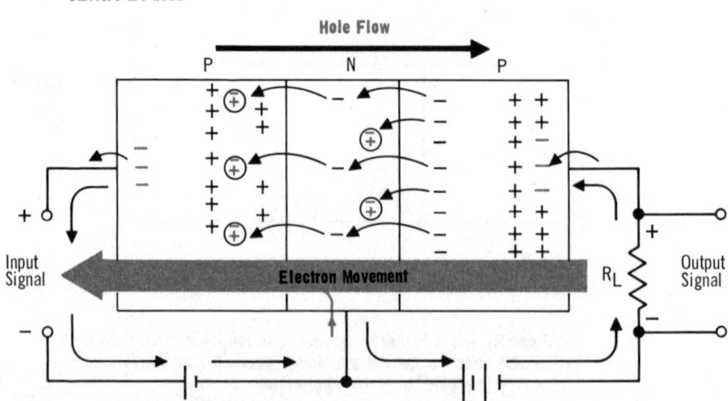

In the p-n-p transistor, collector current changes with emitter current to give gain. Holes, which are the majority carriers, move in the opposite direction to that of the electrons

junction transistor element characteristics

Although the transistor theory presented showed that the current flow through the transistor is similar in many respects to p-n diodes, the transistor has certain basic differences that allow it to work as an amplifier.

First, all three segments of the transistor are doped differently, and second, they are "grown" in different sizes. Since the emitter is forward biased, and provides majority current flow, it is heavily doped to produce a large number of majority carriers. The collector, being reverse biased, provides minority current flow, and so is lightly doped to keep majority carriers low. This is important, since the attractive influence of majority carriers would tend to inhibit the production of minority carriers by thermal agitation. Therefore, the collector can produce a larger number of minority carriers in this way.

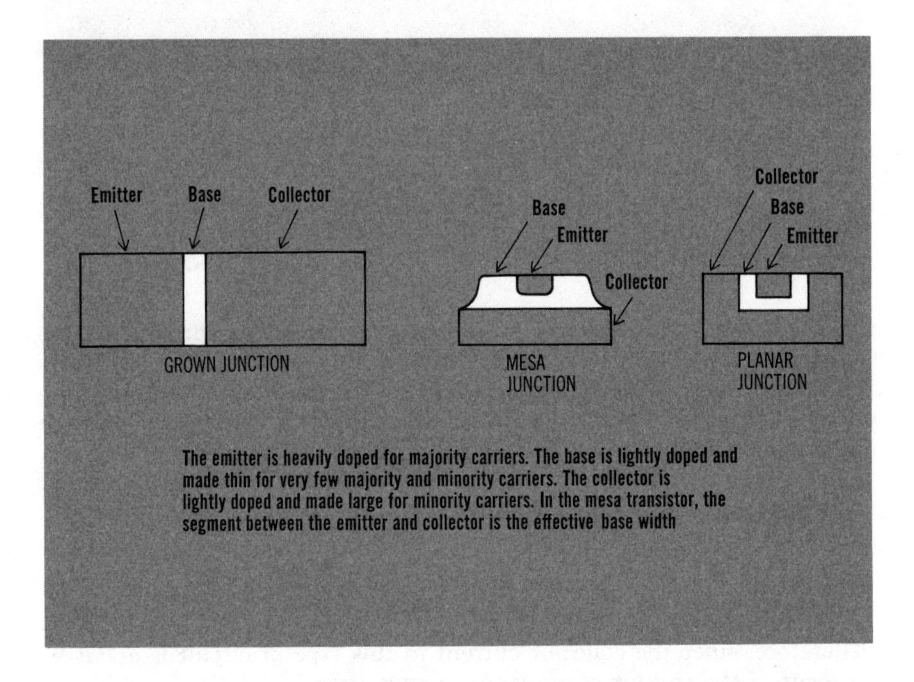

The emitter is heavily doped for majority carriers. The base is lightly doped and made thin for very few majority and minority carriers. The collector is lightly doped and made large for minority carriers. In the mesa transistor, the segment between the emitter and collector is the effective base width

In addition, the collector is made larger than the emitter, so that a sufficient number of minority carriers will be produced with a proper reverse bias so that collector current can be nearly as great as emitter current. The base is very lightly doped to produce very few majority carriers, and is made thin to keep the minority carriers low as well. The thinness of the base is also needed to allow the current carriers to diffuse easily between the emitter and collector.

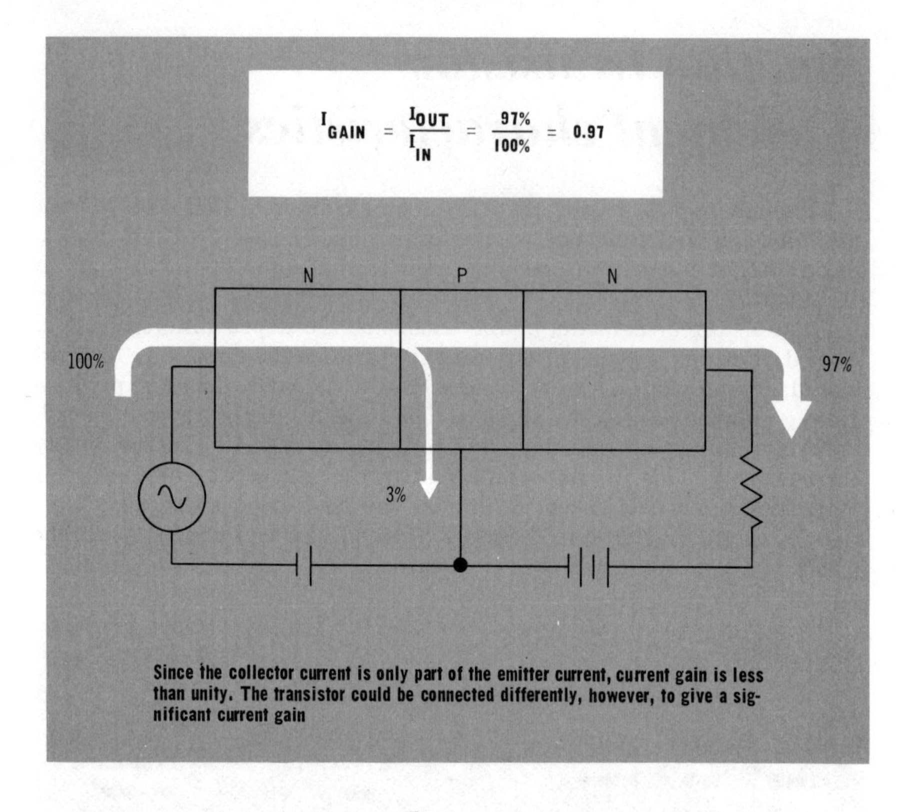

Since the collector current is only part of the emitter current, current gain is less than unity. The transistor could be connected differently, however, to give a significant current gain

current gain characteristics

If you recall, the collector current in a transistor is supplied by the carriers provided by the emitter. Actually, the emitter current follows two paths: one is through the collector, and the other is through the base. Only about 2 to 5 percent of the emitter current goes through the base, while the other 95 to 98 percent becomes the collector current. Since the emitter current is the input current, and the collector current is the output current, the current gain can be computed by:

$$\text{Current gain} = \frac{\text{output current}}{\text{input current}}$$

Therefore, since the collector current in this type of circuit is about 97 percent of the emitter current, the current gain is

$$\text{Current gain} = \frac{97\%}{100\%} = 0.97$$

You can see then that for the type of circuit shown, the current gain is slightly less than unity (a slight loss). However, you will learn that the transistor can be connected in other ways to give an actual current gain.

resistance, voltage, and power gains

Resistance gain is computed similarly to current gain. It is the ratio of the output resistance to the input resistance:

$$\text{Resistance gain} = \frac{\text{output resistance}}{\text{input resistance}}$$

Since the reverse biased base-collector junction provides a high output resistance, and the forward biased base-emitter junction provides a low input resistance, this type of circuit gives a high resistance gain. For example, suppose a transistor is biased in such a way that the emitter-to-base resistance is 150 ohms, and the collector-to-base resistance is 15,000 ohms. The resistance gain would be:

$$\text{Resistance gain} = \frac{\text{output resistance}}{\text{input resistance}} = \frac{15{,}000}{150} = 100$$

The signal voltage gain could be found by dividing the output voltage change for a given input voltage change, if you wanted to take measurements. However, since you know the current gain and the resistance gain, you can find the voltage gain by Ohm's Law:

$$\text{Voltage gain} = I_{GAIN} \times R_{GAIN} = 0.97 \times 100 = 97$$

This means that the output voltage change will be 97 times greater than the input voltage change.

Power gain can be found in a similar way:

$$\text{Power gain} = E_{GAIN} \times I_{GAIN} = 97 \times 0.97 \doteq 94.09$$

or

$$\text{Power gain} = I^2_{GAIN} \times R_{GAIN} = 0.97 \times 0.97 \times 100 = 94.09$$

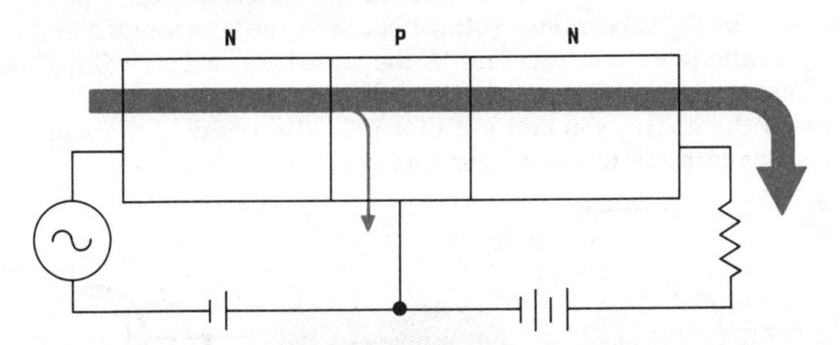

Gain	Equation
Current Gain	$I_{GAIN} = I_{OUT} \div I_{IN} = E_{GAIN} \div R_{GAIN}$
Resistance Gain	$R_{GAIN} = R_{OUT} \div R_{IN} = E_{GAIN} \div I_{GAIN}$
Voltage Gain	$E_{GAIN} = E_{OUT} \div E_{IN} = I_{GAIN} \times R_{GAIN}$
Power Gain	$P_{GAIN} = P_{OUT} \div P_{IN} = E_{GAIN} \times I_{GAIN}$
	$= I^2_{GAIN} \times R_{GAIN} = E^2_{GAIN} \div I_{GAIN}$

comparison of tubes and junction transistors

You can see that although electron tubes and transistors are completely different types of devices, which operate in different ways, they are somewhat related. It might be useful, then, to compare them to get a better idea of how each can be used to do similar jobs.

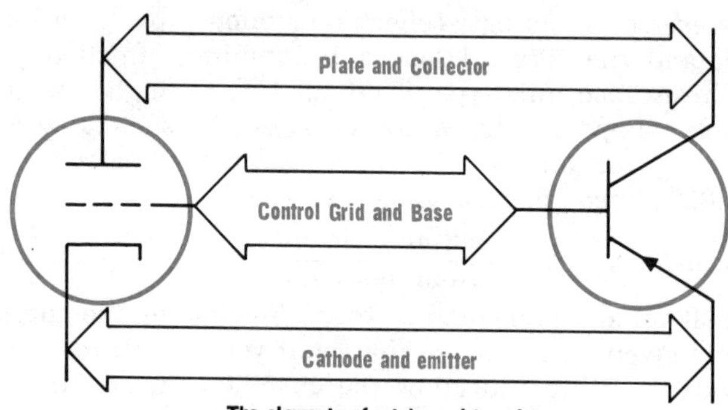

The elements of a tube and transistor perform similar functions

In a tube, the plate *collects* the current, as does the collector in the transistor; so, these two elements can be considered to serve similar functions. In the tube, the current carriers are sent to the plate by the *emission* of the cathode, which is similar to how the emitter supplies carriers to the collector in a transistor. In a tube, the current carriers pass through the control grid; and in the transistor, they pass through the base. In the tube, a bias voltage between the control grid and cathode controls the plate current; and in the transistor, a bias voltage between the base and emitter controls the collector current.

From the above, you can see that it is also possible to compare basic transistor circuits to basic tube circuits.

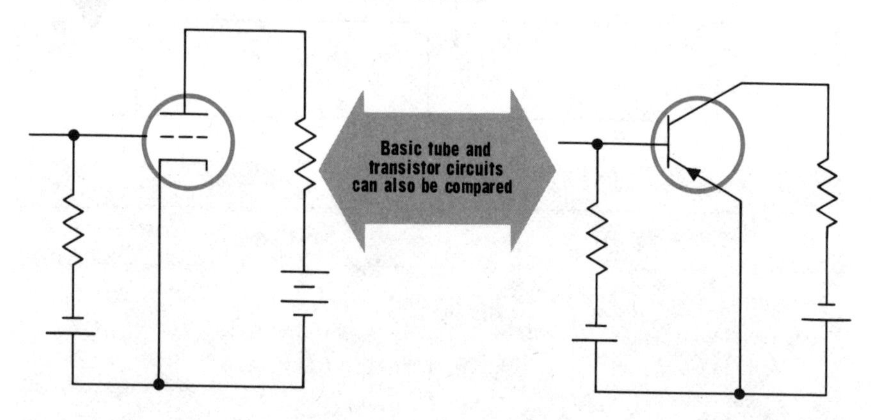

summary

☐ Junction transistors are three-element semiconductors that can amplify a signal. ☐ Junction transistors have an emitter, a base, and a collector. ☐ The emitter supplies the majority carriers for current flow. ☐ The collector collects the emitter's majority carriers for circuit operation. ☐ The base controls the current flow between the emitter and the collector. ☐ The emitter is represented by an arrow that points in the direction of hole flow. ☐ The arrow points toward the base for a p-type emitter and away from the base for an n-type emitter.

☐ The base in a transistor is common to both the emitter and collector circuits. ☐ Since the emitter-base circuit is forward biased, majority carriers cross the junction from the emitter into the base. The base provides few majority carriers to combine with the incoming emitter majority carriers. ☐ The collector-base circuit is reverse biased, and minority carriers in the collector are attracted to the base junction. The base provides few minority carriers to combine with the collector minority carriers. ☐ The excess majority carriers from the emitter pass through the base and are attracted to the collector where they combine with the minority carriers. ☐ The entire emitter current (less the 2 to 5% that flows down the base) reaches the collector.

☐ Current gain of a transistor $= I_{OUT} \div I_{IN} = E_{GAIN} \div R_{GAIN}$. Current gain in a circuit with a common base is less than 1 due to the input current that flows in the base. ☐ Resistance gain $= R_{OUT} \div R_{IN} = E_{GAIN} \div I_{GAIN}$. The reverse-biased collector-base output circuit provides a high resistance compared with the forward-biased emitter-base input circuit; typically, it is 100 times greater ☐ Voltage gain $= E_{OUT} \div E_{IN} = I_{GAIN} \times R_{GAIN}$. The resistance of the output circuit provides voltage gain although the current gain is less than 1. ☐ Power gain $= P_{OUT} \div P_{IN} = E_{GAIN} \times I_{GAIN}$. In a circuit with a common base, power gain is slightly less than voltage gain due to lack of current gain.

review questions

1. Describe the function of each element in a junction transistor?
2. Draw the symbol for a p-n-p junction transistor.
3. Draw the symbol for an n-p-n junction transistor.
4. Why is the emitter-base circuit forward biased?
5. Why is the collector reverse biased?
6. How is voltage gain provided in a transistor?
7. Write the equations for current gain.
8. Write the resistance and power gain equations.
9. Explain why the current gain in a transistor with a common-base circuit is always less than 1.
10. Why is the junction transistor called bipolar?

the common-base circuit

As shown, the common-base circuit is the type you have been studying all along. It is so called because the base is common to both the input and output. Since the base in a transistor has a similar function to the control grid in a tube, the common-base circuit is similar to the common-, or grounded-, grid amplifier.

In the common-base amplifier, the input signal is applied to the emitter, and the output signal appears across R_L at the collector. When the input signal aids the emitter-base bias, the emitter current, I_E, goes up. This causes the collector current, I_C, to go up also, producing a larger voltage drop across the load resistor. Since the change in collector current is almost equal to the change in emitter current, and the output resistance is much greater than the input resistance, the voltage change across the load resistor is greater than the input signal swing. Thus, the signal is amplified.

By the same token, since the change in input current is about the same as the change in output current, but the output voltage swing is greater, the output power is greater than the input power. This is explained more fully later.

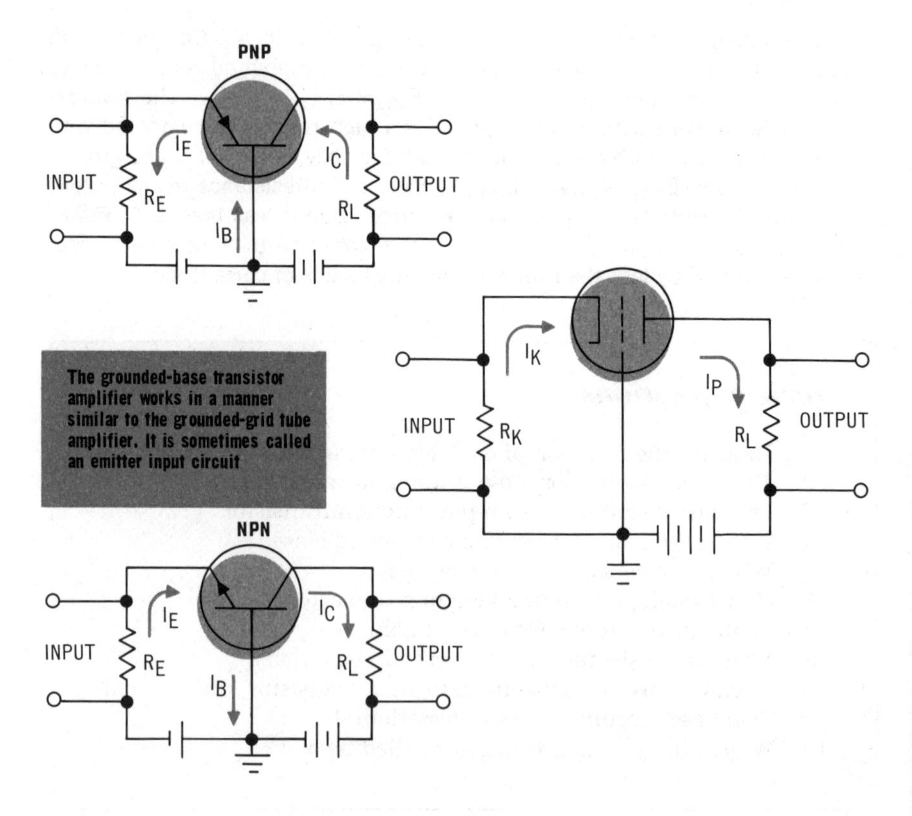

The grounded-base transistor amplifier works in a manner similar to the grounded-grid tube amplifier. It is sometimes called an emitter input circuit

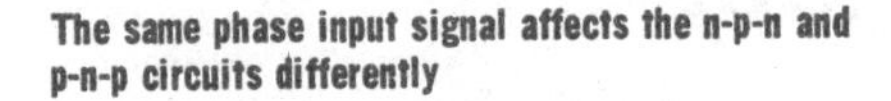
The same phase input signal affects the n-p-n and p-n-p circuits differently

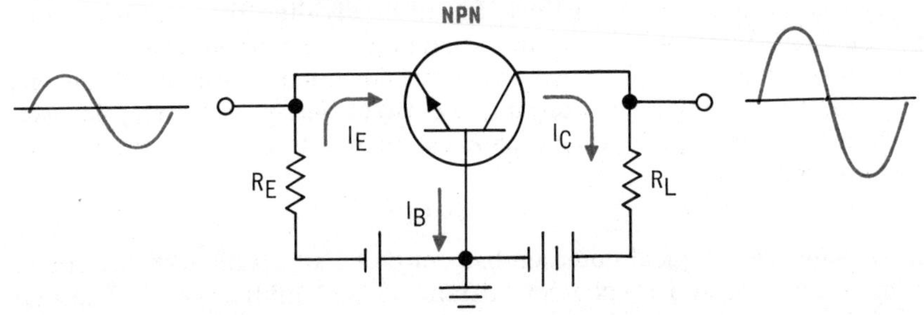

N-p-n circuits and p-n-p circuits are affected differently by the same phase input signal because they have opposite-polarity emitter bias

But, since they also have opposite-polarity collector bias, the output phases of both are the same. There is no phase reversal in a common-base circuit

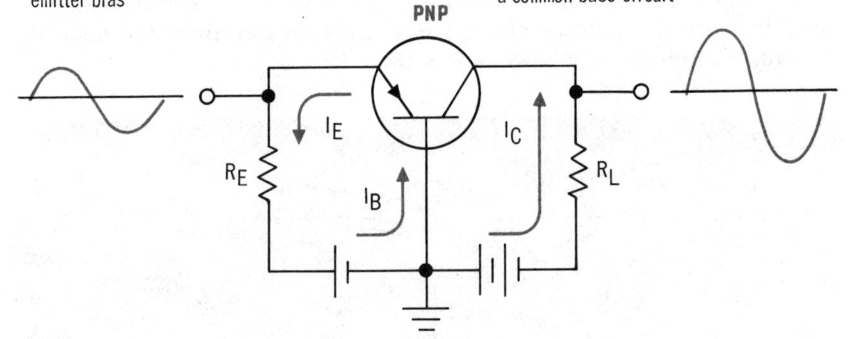

input and output phases

From Volume 3, you should recall that the grounded-grid tube amplifier did *not* reverse the phase of the input signal. Neither does the common-base amplifier. In the n-p-n circuit, when the input signal goes positive, it *subtracts* from the emitter bias, reducing the overall bias. This reduces the emitter current and the collector current as well. With less collector current, the drop across the load resistor is less. Therefore, the voltage at the collector, which is the output voltage, goes more positive. The output voltage, then, goes in the same direction as the input voltage, and has the same phase.

With the p-n-p circuit, when the input signal goes positive, it *aids* the emitter bias, and causes more emitter and collector current to flow. The increased collector current causes a greater drop across the load resistor. As a result, the collector voltage becomes less negative, or swings in the positive direction, following the phase of the input signal.

Even though the same signal affects the n-p-n and p-n-p currents differently, the output phases are the same because of the opposite bias polarities.

current gain

As you learned, current gain is found by dividing the output current by the input current. The output current in a common-base circuit is the collector current (I_C), and the input current is the emitter current (I_E). In the common-base circuit, the Greek letter *alpha* (α) is used to signify current gain. Using symbols, then, the equation for current gain becomes:

$$\alpha = I_C/I_E$$

Now, suppose a typical common-base circuit had an emitter current of 6 milliamperes, and a collector current of 5.83 milliamperes. The current gain would be

$$\alpha = I_C/I_E = 5.83/6 = 0.97$$

As you know, the current gain in this type of circuit has to be less than unity because the emitter current is always greater than the collector current. Therefore, α is always less than 1.

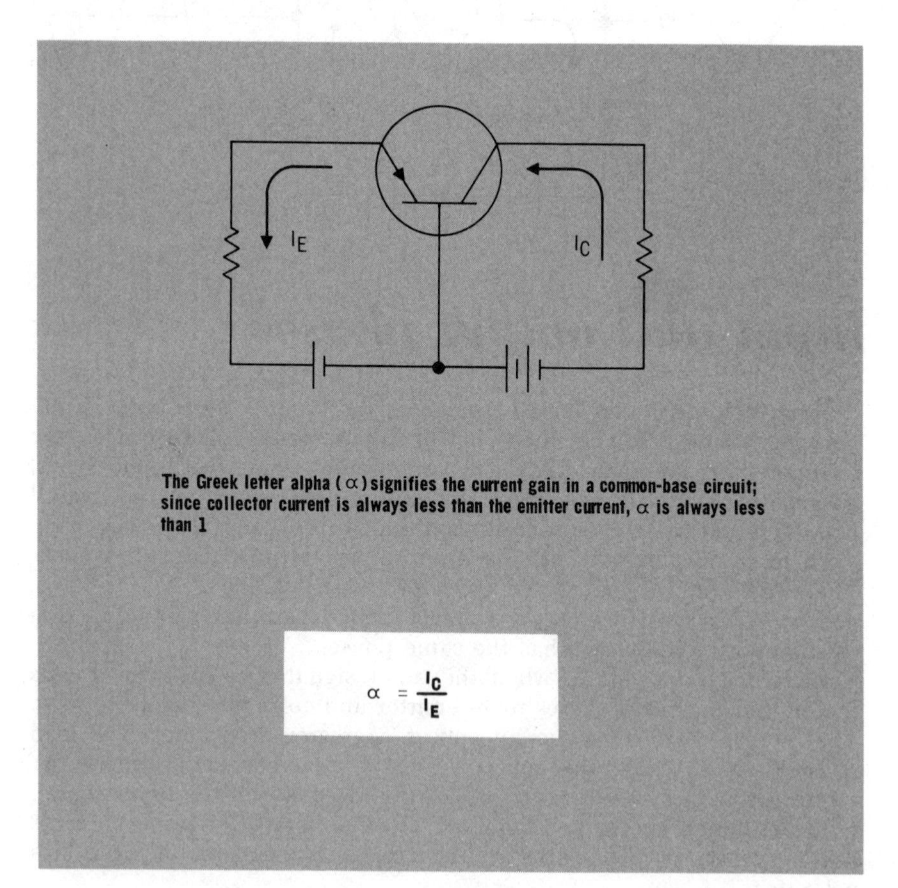

The Greek letter alpha (α) signifies the current gain in a common-base circuit; since collector current is always less than the emitter current, α is always less than 1

resistance, voltage, and power gains

For the common-base circuit, even though it does not have a true current gain, it can give voltage and power gains because of its resistance gain. The resistance gain is brought about because the collector circuit is reverse biased, and the input circuit is forward biased; thus, the output circuit has a much higher resistance than the input circuit.

Actually, it is the output load resistor (R_L) that determines the output resistance, since it produces the varying output voltage as the collector current changes. If this load resistor could be made equal to the collector-base junction resistance, the maximum gain could be realized. But this resistance could be up in the megohm range, and such a higher value would require a very large bias battery to get the proper bias current to flow. So, the load resistance is usually much less. A typical circuit could have an input resistance of 300 ohms, and an output resistance of 15,000 ohms. The resistance gain of such a circuit is:

$$R_{GAIN} = R_{OUT}/R_{IN} = 15{,}000/300 = 50$$

By Ohm's Law, then, if this circuit had a current gain of 0.97, the voltage and power gains are

$$E_{GAIN} = \alpha \times R_{GAIN} = 0.97 \times 50 = 48.5$$
$$P_{GAIN} = E_{GAIN} \times \alpha = 48.5 \times 0.97 = 47.04$$

Of course, both voltage and power gains could also be found by:

$$E_{GAIN} = E_{OUT}/E_{IN} \qquad P_{GAIN} = P_{OUT}/P_{IN}$$

Gain	Equation
Current Gain	$\alpha = \frac{I_C}{I_E}$
Resistance Gain	$R_{GAIN} = \frac{R_{OUT}}{R_{IN}}$
Voltage Gain	$E_{GAIN} = \alpha \times R_{GAIN}$
	$E_{GAIN} = \frac{E_{OUT}}{E_{IN}}$
Power Gain	$P_{GAIN} = E_{GAIN} \times \alpha$
	$P_{GAIN} = \alpha^2 \times R_{GAIN}$
	$P_{GAIN} = \frac{E^2_{GAIN}}{R_{GAIN}}$
	$P_{GAIN} = \frac{P_{OUT}}{P_{IN}}$

the common-emitter circuit

In the *common-emitter* circuit, the emitter is common to both the input and output. Since the function of the emitter is similar to that of a cathode, the common-emitter transistor amplifier is similar to the common-, or grounded-, cathode tube amplifier. And like its tube counterpart, the grounded-emitter amplifier is the type of circuit used most often. It is sometimes called a *base input* circuit.

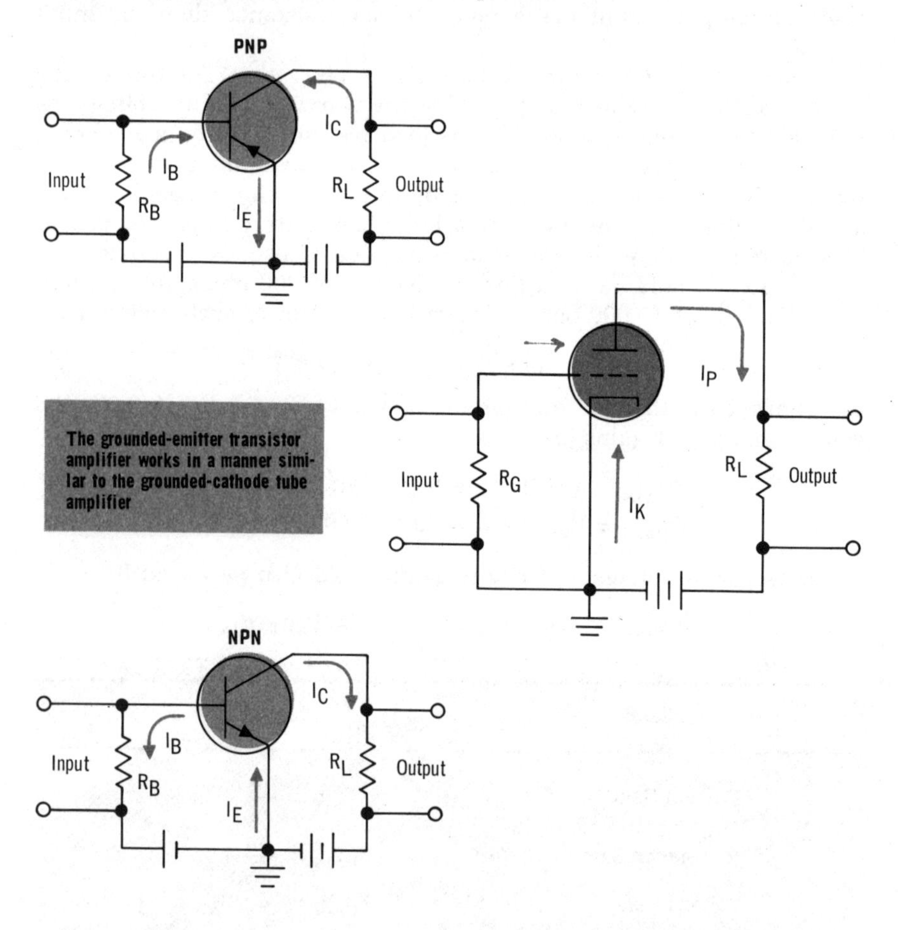

The grounded-emitter transistor amplifier works in a manner similar to the grounded-cathode tube amplifier

The input signal in this type of circuit is applied to the base, and the output signal appears across R_L at the collector. The input signal either aids or opposes the base-emitter bias battery. When it aids the bias battery, the base current, I_B, goes up, because the emitter current, I_E, is increased due to the higher forward bias. Therefore, the collector current, I_C, goes up, too, to increase the drop across the load resistor. The opposite happens when the input signal goes down. The output voltage swing is greater than the input signal to give voltage gain.

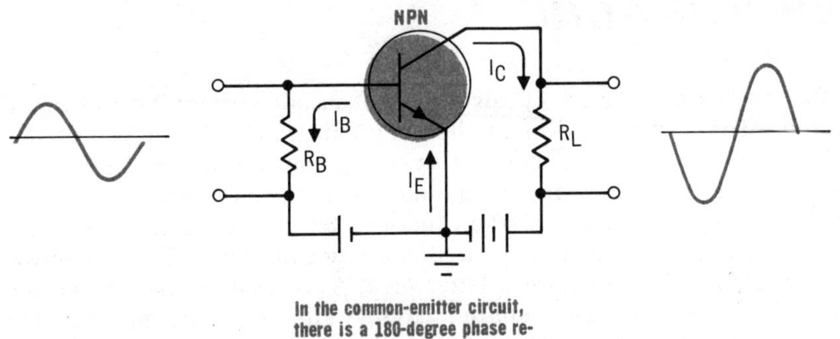

In the common-emitter circuit, there is a 180-degree phase reversal of the signal voltage

input and output phases

From the previous description, you may have the idea that the grounded-emitter amplifier works just like the grounded-base amplifier. There is one big difference, however. In Volume 3, you learned that the phase of a signal is *reversed* in a grounded-cathode tube; this also happens in a grounded-emitter transistor. The reason for this is that in the grounded-emitter circuit, the input signal is applied from base to emitter, whereas in the grounded-base circuit, the signal is applied from emitter to base. As a result, the same phase input signal affects the base-emitter bias in both circuits in opposite ways.

For example, in the n-p-n common-emitter circuit, when the input signal goes positive, it *adds* to the base-emitter bias, increasing emitter and base currents. It also increases the collector current to cause a greater drop across the load resistor. As a result, the voltage at the collector becomes less positive, or swings in a negative direction. The output voltage, then swings *opposite* to the input voltage.

With the p-n-p circuit, a positive input voltage *opposes* base-emitter bias, lowering the base, emitter, and collector currents. This reduces the drop across the load resistor, so that the collector voltage becomes more negative, which again is opposite to the input signal phase.

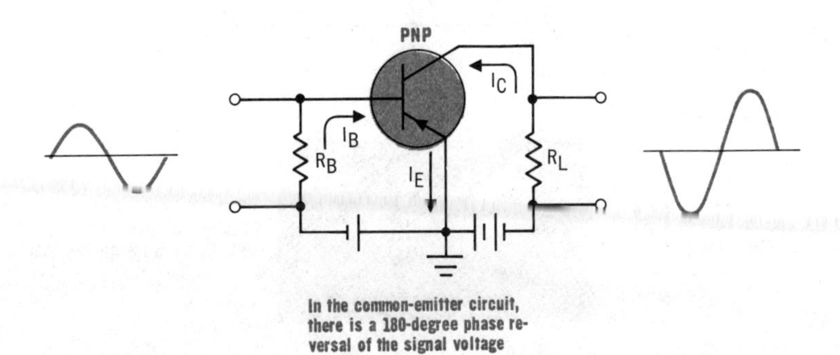

In the common-emitter circuit, there is a 180-degree phase reversal of the signal voltage

current gain

Current gain is found by dividing the output current by the input current. The output current in the common-emitter circuit is still the collector current (I_C). The input current, however, is no longer the emitter current, as it is in the common-base circuit; instead, it is the base current (I_B). Therefore, the equations for current gain in the common-base and common-emitter circuits are not the same. To distinguish one from the other, the Greek letter *beta*, β, is used as the symbol for current gain in a common-emitter circuit. Therefore, the equation for current gain becomes:

$$\beta = I_C/I_B$$

Remember, the current gain in the common-base circuit was less than unity because the input current (I_E) was greater than the output current (I_C). However, since the input current in the common-emitter circuit is the base current (I_B) the output current is much greater than the input current; and so the circuit has a high current gain. For example, suppose the circuit shown has a collector current of 5.83 milliamperes, and an emitter current of 6 milliamperes; find the current gain. As you can see, first you have to find the base current. Since the emitter current is the sum of the base and collector currents, the base current can be found by:

$$I_E = I_C + I_B \quad \text{and} \quad I_B = I_E - I_C = 6 - 5.83 = 0.17 \text{ ma}$$

Now you can find current gain by:

$$\beta = I_C/I_B = 5.83/0.17 = 34.29$$

The high current gain of a common-emitter circuit is the property that makes this circuit much more useful than the common-base circuit.

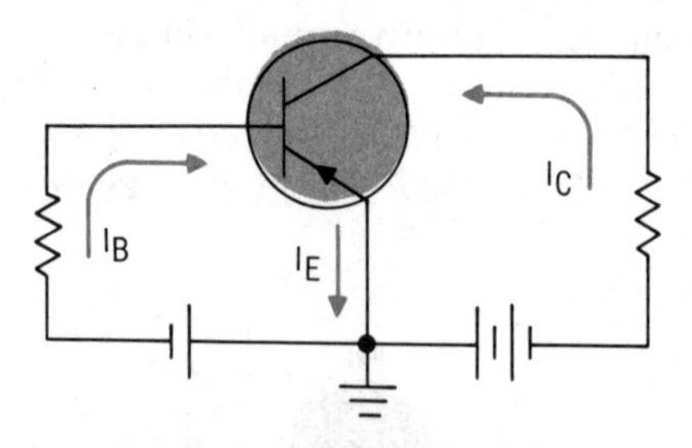

The Greek letter beta (β) signifies the current gain in a grounded-emitter circuit

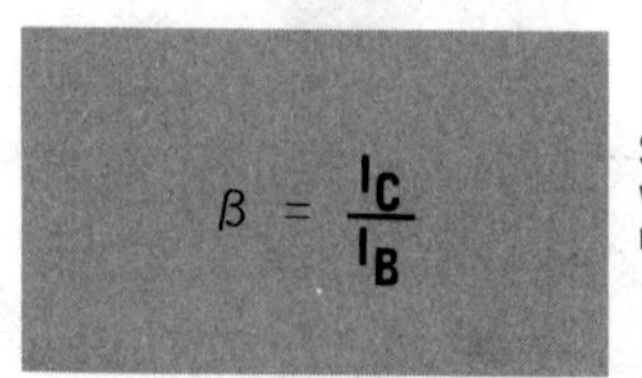

Since the collector current is always greater than the base current, β is always greater than 1

resistance, voltage, and power gains

Since the grounded-emitter amplifier has high current gain, it is capable of giving much higher voltage and power gains than the grounded-base circuits. This is particularly true since the resistance gain of the grounded-emitter circuit is just as high as in the grounded-base circuit; remember, the input and output junctions, and, therefore, resistances are the same in both circuits.

Suppose a circuit had an input resistance of 500 ohms, an output resistance of 5000 ohms, and a β of 35. The resistance gain, then, is

$$R_{GAIN} = 5000/500$$
$$= 10$$

Using Ohm's Law, then, since β is 35, the voltage gain is

$$E_{GAIN} = R_{GAIN} \times \beta$$
$$= 10 \times 35$$
$$= 350$$

The power gain is

$$P_{GAIN} = E_{GAIN} \times \beta$$
$$= 350 \times 35$$
$$= 12{,}250$$

As with the grounded-base circuit, both voltage and power gains could also be found by:

$$E_{GAIN} = E_{OUT}/E_{IN}$$

and

$$P_{GAIN} = P_{OUT}/P_{IN}$$

Gain	Equation
Current Gain	$\beta = I_C/I_B$
Resistance Gain	$R_{GAIN} = R_{OUT} \div R_{IN}$
Voltage Gain	$E_{GAIN} = \beta \times R_{GAIN}$
	$E_{GAIN} = E_{OUT} \div E_{IN}$
Power Gain	$P_{GAIN} = E_{GAIN} \times \beta$
	$P_{GAIN} = \beta^2 \times R_{GAIN}$
	$P_{GAIN} = E^2_{GAIN} \div R_{GAIN}$
	$P_{GAIN} = P_{OUT} \div P_{IN}$

the common-collector circuit

In the *common-collector* circuit, the collector is common to both the input and output. Since the function of the collector is similar to that of a tube's plate, the common-collector transistor circuit is similar to the common-, or grounded-, plate amplifier. And, like its tube counterpart, if the collector is not actually at d-c ground, it must be kept at a-c ground with a bypass capacitor. In the tube circuit, since the output voltage is taken off the cathode, the circuit is often called a *cathode follower*. For the same reason, the transistor circuit is also called an *emitter follower*.

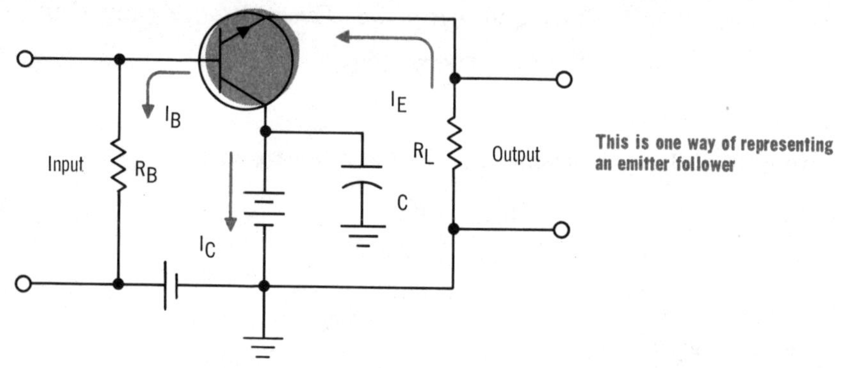

This is one way of representing an emitter follower

The input signal to this circuit is applied to the base, and the output signal is taken off across R_L in the emitter. As in the other circuits, the input signals either aid or oppose the input bias voltage, but unlike the other circuits there is no gain, as you will learn. Essentially, the collector circuit performs no active function, and the input signal merely causes the emitter current to rise and fall.

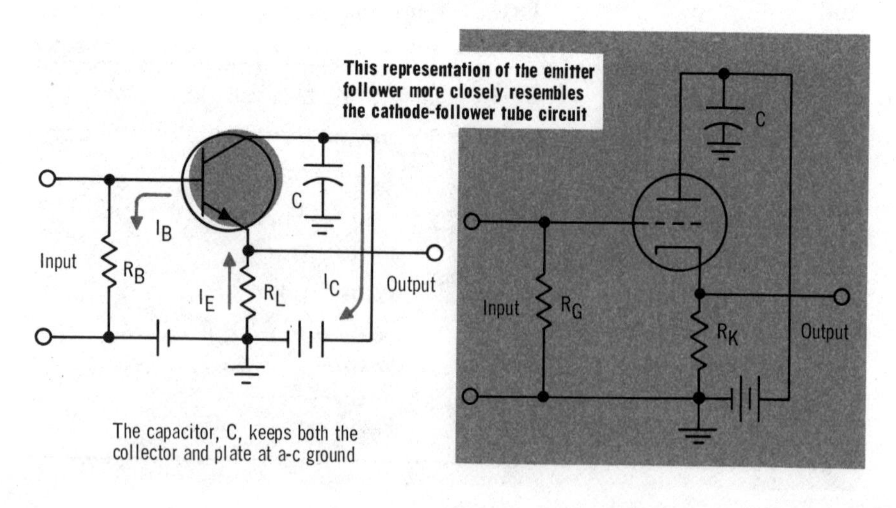

This representation of the emitter follower more closely resembles the cathode-follower tube circuit

The capacitor, C, keeps both the collector and plate at a-c ground

input and output phases

You can see now that the emitter follower operates on the same principle as the other circuits. The input signal either aids or opposes the base-emitter bias voltage to increase or decrease the emitter current. However, the effect that this has on the collector current is not significant. Essentially, the collector is connected in the circuit to allow emitter current flow. Remember, as you learned earlier, that without the collector to pick up current carriers, the carriers from the emitter would accumulate in the base to restrict emitter-current flow. So the collector is merely used to complete the d-c current path and is kept at a-c zero so that it does not carry signal current.

The circuit configuration of the emitter follower on this page shows how simply the circuit works. In the n-p-n transistor, when the input signal swings positive, it *aids* the bias to increase emitter current. The increased emitter current raises the voltage drop across the load resistor, R_L, to make the output voltage more positive; so the input and output signals are *in phase*.

With the p-n-p transistor, a positive input signal *opposes* the bias to reduce emitter current. This lowers the voltage drop across R_L. But, because of the direction of current flow, the emitter becomes less negative, or swings in a positive direction; the input and output signals, then, are in phase.

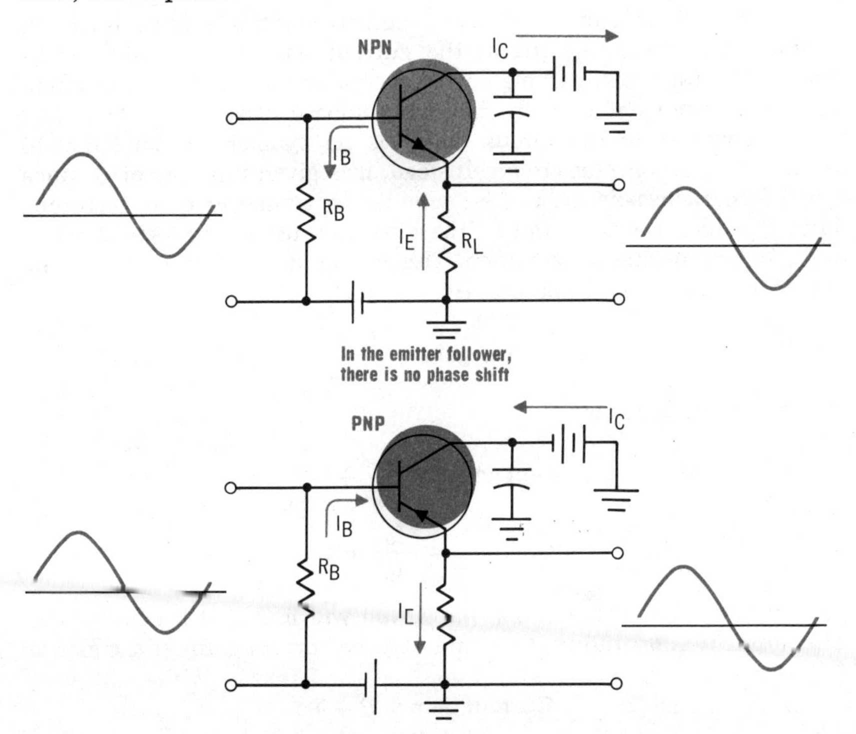

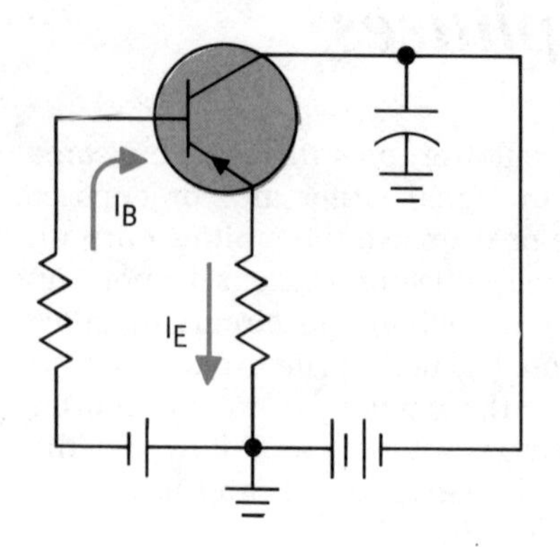

Current gain for the common-emitter circuit is found in terms of β

Current Gain $= \beta + 1$

$= \dfrac{I_E}{I_B}$

current gain

Current gain in the emitter follower is found by dividing the output current by the input current, just as with the other circuits. However, the input current here is the base current (I_B), and the output current is the emitter current (I_E), so the equation is

$$\text{Current gain} = I_E/I_B \qquad (1)$$

Since the emitter current in the common-collector circuit is much greater than the base current, the current gain is very high, even higher than for a comparable common-emitter circuit. Although alpha (α) denotes current gain in the common-base circuit, and beta (β) for the common-emitter circuit, there is no symbol for current gain in the common-collector circuit. Instead, it is given in terms of β, since β is the characteristic most often supplied by transistor manufacturers. Since β is found with I_C and I_B, these terms must be substituted for I_E in the common-collector equation. This can be done since $I_E = I_C + I_B$. Therefore, equation (1) becomes:

$$\text{Current gain} = \frac{I_C + I_B}{I_B}$$

which can also be shown as:

$$\text{Current gain} = \frac{I_C}{I_B} + \frac{I_B}{I_B}$$

or

$$\text{Current gain} = \frac{I_C}{I_B} + 1$$

Also, if you check the equation for β, you will find that it equals I_C/I_B. Therefore, by substituting β, we find that the current gain for the emitter follower is

$$\text{Current gain} = \beta + 1$$

resistance, voltage, and power gains

You might think that since the emitter follower has a higher current gain than even the common-emitter circuit that it would also have higher voltage and power gains. It does not because the circuit has a resistance gain *much lower* than unity. You can see this when you realize that the input circuit is the base-collector, which is reverse biased, and the output is the emitter-collector, which is forward biased. Therefore, using the resistance gain equation, you can see that the gain is very low:

$$R_{GAIN} = R_{OUT}/R_{IN} = R_{LOW}/R_{HIGH} = \text{much less than } 1$$

Actually, a typical input resistance for this circuit is about 300K, and a typical output resistance is about 300 ohms. So, you can see that the resistance gain is very low. It is so low, in fact, that even though the circuit has a high current gain, the voltage gain that results from multiplying the current and resistance gains amounts to a loss. For example, using the resistance figures given above with a current gain of 800, the voltage gain would be

$$E_{GAIN} = R_{GAIN} \times (\beta + 1) = 0.001 \times 800 = 0.8$$

Power gain, though, can be greater than 1 because it is found by squaring the current gain:

$$P_{GAIN} = (\beta + 1)^2 \times R_{GAIN}$$

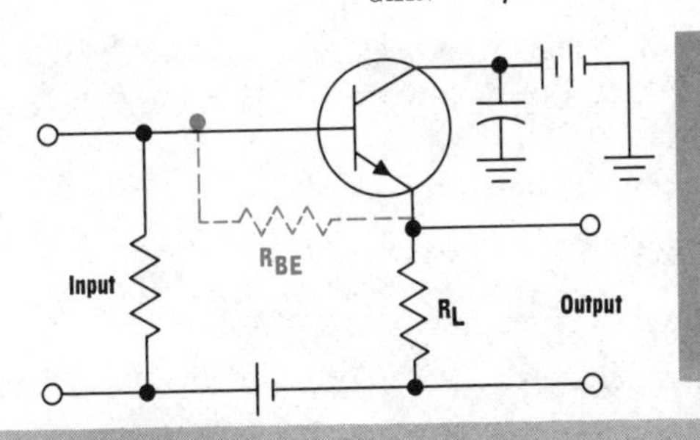

This circuit shows why there is always a voltage loss. R_{BE} represents the base-emitter junction resistance. The input signal is actually applied across R_{BE} in series with the load resistor, R_L. Therefore, only part of the input voltage can be dropped across R_L; the output voltage, then, must always be less than the input voltage

Gain	Equation
Current Gain	$I_{GAIN} = \beta + 1$
Resistance Gain	$R_{GAIN} = R_{OUT} \div R_{IN}$
Voltage Gain	$E_{GAIN} = R_{GAIN} \times (\beta + 1)$
	$E_{GAIN} = E_{OUT} \div E_{IN}$
Power Gain	$P_{GAIN} = E_{GAIN} \times (\beta + 1)$
	$P_{GAIN} = (\beta + 1)^2 \times R_{GAIN}$
	$P_{GAIN} = P_{OUT} \div P_{IN}$

alpha and beta

Transistors are manufactured for certain values of α and β. But, as mentioned earlier, the common-emitter circuit is the one that is used most often. As a result, transistor manufacturers usually only give the β characteristic for the transistor. However, the common-base circuit is still used, and so it is important also to know α. So, if only β is known, there should be some way of finding α without making too many measurements. The same is true if you only know α, and β is needed. Fortunately, there is a definite relationship between the two, so that each can be expressed in terms of the others. These equations are

$$\beta = \frac{\alpha}{1-\alpha}$$

and

$$\alpha = \frac{\beta}{1+\beta}$$

These equations were derived by substituting expressions in the basic equation for α and β to find each in terms of the other. So, if you know one, you can find the other.

Solving for β in terms of α

1. Since $\alpha = I_C/I_E$, solving for I_C, we have: $I_C = \alpha I_E$.
2. Therefore, since $\beta = I_C/I_B$, then by substituting for I_C, $\beta = \alpha I_E/I_B$.
3. Also, $I_B = I_E - I_C$. And solving for I_C, we have: $I_C = I_E - I_B$.
4. Therefore, substituting for I_C found in step 1, $I_E - I_B = \alpha I_E$.
5. Then by transposing all the terms with I_E to the left, $I_E - \alpha I_E = I_B$, which is simplified to $I_E(1 - \alpha) = I_B$.
6. Now the β found in step 2 can become $\beta = \dfrac{\alpha I_E}{I_E(1-\alpha)}$
7. Factoring the I_E, the equation is simplified to:

$$\beta = \frac{\alpha}{1-\alpha}$$

Solving for α in terms of β

1. Since $\beta = \alpha/(1 - \alpha)$, then by transposing, $\alpha = \beta - \beta\alpha$.
2. Transposing further, $\alpha + \beta\alpha = \beta$.
3. Simplifying, $\alpha(1 + \beta) = \beta$.
4. Solving for α, we have

$$\alpha = \frac{\beta}{1+\beta}$$

summary

□ No phase reversal occurs between the input and output signals in the common-base circuit. □ Alpha (α) signifies current gain in the common-base circuit and equals I_C/I_E. □ Alpha (α) is always less than 1 since the collector current is always less than the emitter current because of current flow in the base. □ The common-base circuit provides voltage and power gains due to the resistance gain of the reverse-biased output circuit compared with the forward-biased input circuit.

□ Phase reversal occurs between the input and output signals in the common-emitter circuit. □ Beta (β) signifies current gain in the common-emitter circuit, and equals I_C/I_B. □ Beta (β) is always much greater than 1 since the collector current is always greater than the base current. □ The common-emitter circuit provides the highest voltage and power gains due to the high current gain and resistance gain.

□ The output of the common-collector circuit is taken from the emitter; this circuit is also called an emitter follower. □ No phase reversal occurs between the input and output signals in the common-collector circuit. □ Current gain in the common-collector circuit is given in terms of β and equals $\beta + 1 = (I_C + I_B)/I_B$. □ The common-collector circuit provides the highest current gain since the base current is only a small part of the emitter current. □ The common-collector circuit provides no voltage gain and low power gain due to low resistance gain. □ α and β are related and can be expressed in terms of one another: $\alpha = \beta/(1 + \beta)$ and $\beta = \alpha/(l - \alpha)$.

review questions

1. Why is the common-base circuit also called an emitter input circuit?
2. What does α signify?
3. How is α calculated?
4. How does the common-base circuit provide voltage and power gains if there is no current gain?
5. Why is there phase reversal in the common-emitter circuit, but not in the common-base circuit?
6. What does β signify? How is it calculated?
7. Why is the common-collector circuit also called an emitter follower?
8. The current gain for a common-collector circuit can be expressed in terms of β. Why is this possible?
9. Explain why the voltage gain of a common-collector circuit must always be less than 1.
10. Express α and β in terms of each other.

junction transistor characteristics

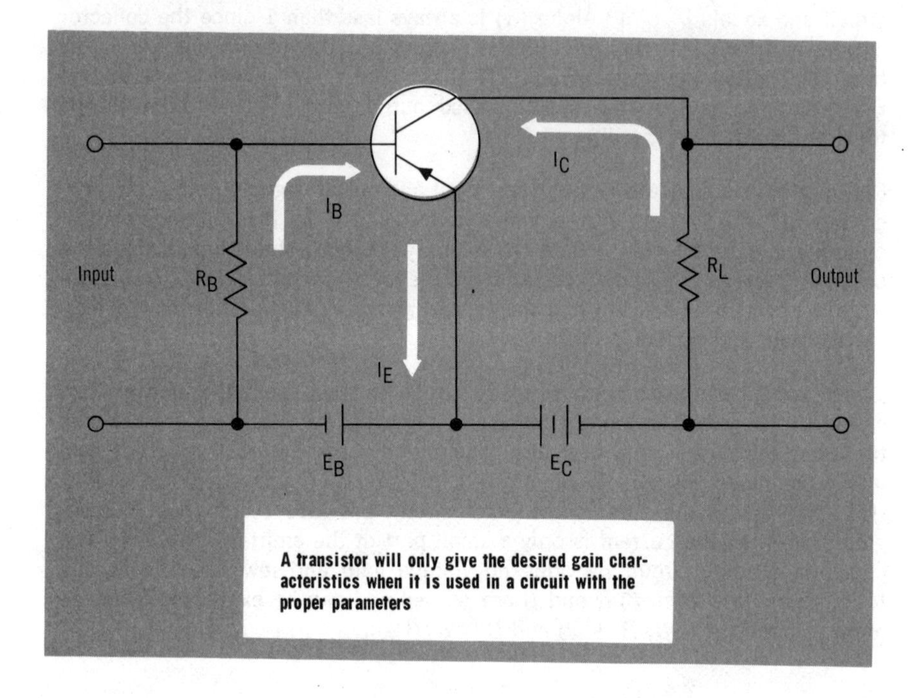

A transistor will only give the desired gain characteristics when it is used in a circuit with the proper parameters

As you learned, junction transistors are manufactured to have specific characteristics that will give them certain values of alpha and beta. However, these current gain values will not be produced if the transistor is connected in a circuit that does not have the proper *parameters*. In other words, the circuit must provide certain specified values of bias voltages and bias currents, as well as certain load resistances, to bring about the characteristics that the transistor was designed to have. As with electron tubes, there are many different types of transistors, each providing certain desired characteristics when they are used in the proper circuits. Basically, the circuit must be designed to certain values of:

1. Emitter-base bias voltage.
2. Collector-base bias voltage.
3. Emitter bias current.
4. Base current.
5. Collector bias current.
6. Input resistance.
7. Output load resistance.

operating curves

As with electron tubes, junction transistors are manufactured to have certain characteristics under various operating conditions. The transistor does not have to be used for only one set of circuit parameters, but can be operated with different combinations of bias voltages and currents to give the desired performance. The various possible combinations of parameters are indicated by a family of *operating curves*, which is provided by the manufacturer for each transistor type. Such a set of characteristic curves is shown.

Each characteristic curve represents a specific value of base current that can be used, and with the curve chosen, you can find an operating point that will give the needed collector current that will flow for a certain collector voltage. You probably recall that collector current is actually determined by emitter current and voltage. However, the base current is also determined by emitter current, and so can be used to indicate it. It is a standard established by transistor manufacturers to plot the curves around base current, since the base input, or common-emitter circuit is used the most. Remember, though, that a measurement of base current is essentially also an indication of how much emitter current is flowing.

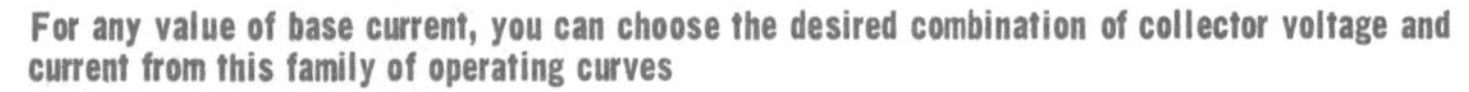
For any value of base current, you can choose the desired combination of collector voltage and current from this family of operating curves

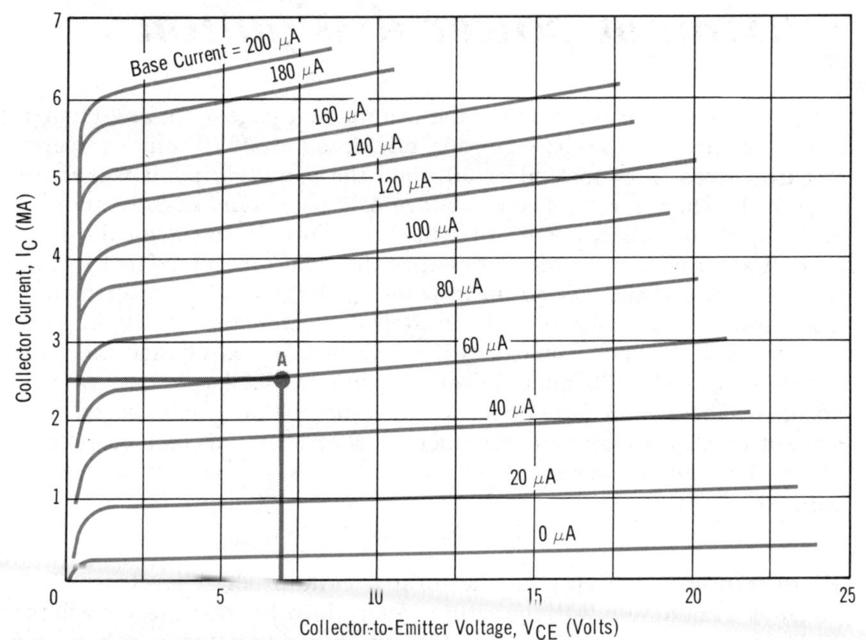

Operating point A shows that if the circuit causes a 60-μA base current, with a bias voltage of 7 volts from the collector to the emitter, the circuit will provide about 2.5 milliamperes of collector current

All operating points must be kept to the LEFT of the maximum power dissipation curve, or the transistor will burn out

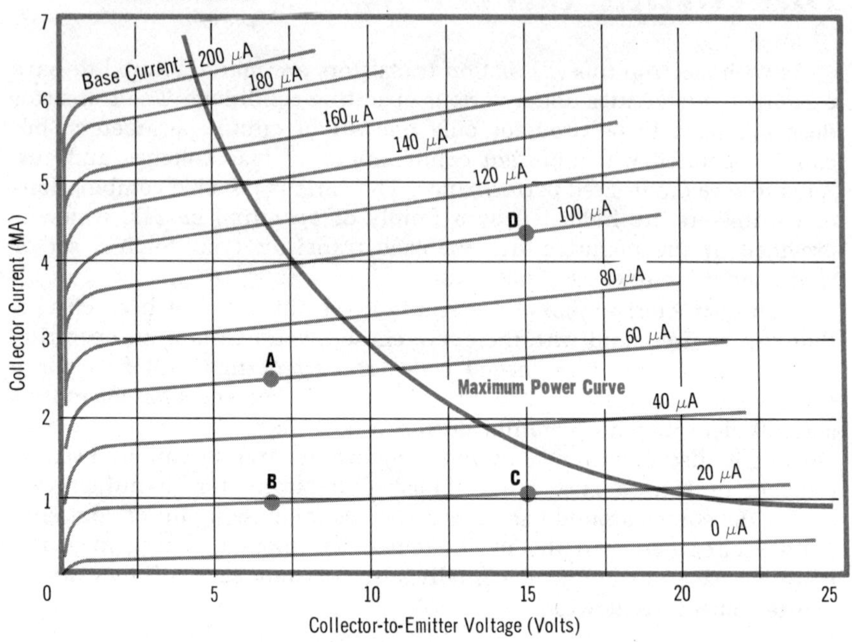

maximum power dissipation

If we take the circuit used for the example on page 4-75, and change the resistance of the base-emitter circuit so that 20 microamperes of base current flow instead of 60, then the *operating point* will move to point B. Then, for a collector voltage of 7 volts, only slightly under 1 milliampere of collector current will flow. Now, if you keep the base current at 20 microamperes, but change the collector voltage to 15 volts, the collector current will go up to point C, slightly over 1 milliampere. This shows that the collector voltage affects collector current only slightly.

Now, with 15 volts applied to the collector, if you change the base circuit to provide 100 microamperes (point D), collector current will go up to about 4.33 milliamperes. This shows that a *change* in base current of 80 *micro*amperes produces a *change* in collector current of about 3.25 *milli*amperes (3250 microamperes), which is a current gain of:

$$\beta = 3250/80 \cong 40$$

Every transistor can only handle up to a certain power level, which is given as a maximum wattage rating, determined by the collector voltage and current (P = EI). Therefore, certain operating points can burn out the transistor. When a maximum power dissipation curve is given as shown on this page, any operating point to the *right* of the curve, such as point D, will exceed the safe wattage rating of the transistor.

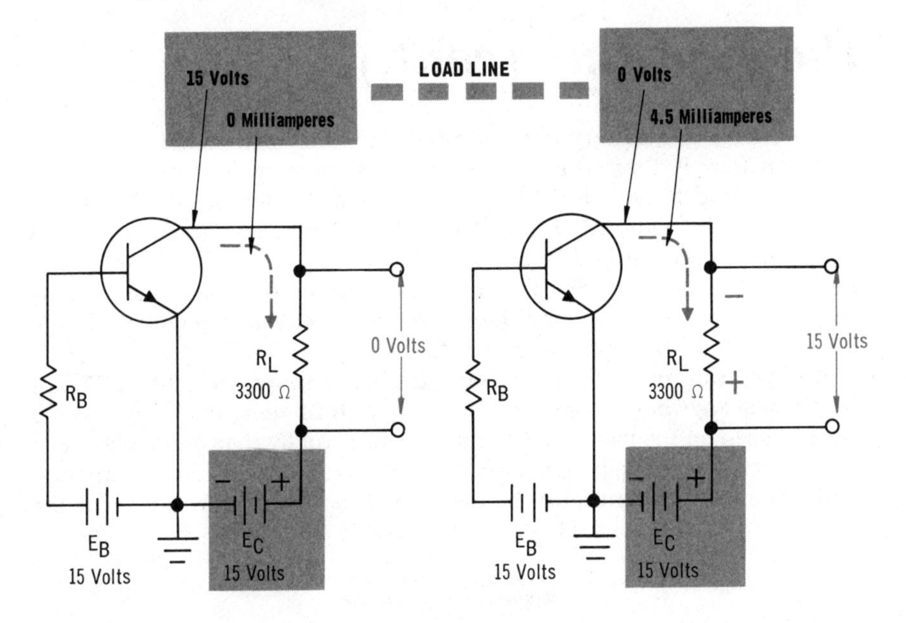

the load line

The manner by which you determine an operating point depends on the parameters with which you start. You can arbitrarily pick your own parameters, but, in many cases, you may have a particular power source or a particular transistor available that you must use. Sometimes, you must use a specific load, while other times you will have some leeway in selecting the load. In some cases, you may be limited to a certain level input signal, and at other times, you might have some control over the input signal. Many times, you can use the "typical values" supplied by the manufacturers, and other times these might not suit your application. All of these conditions affect the way you design a circuit.

For our example, assume that you must use a 3300-ohm load, that you only have two 15-volt batteries, and that the transistor has the proper β characteristic to give enough gain.

The first thing you must do is to draw a *load line,* which will show, with the voltages you have, all of the possible operating points you can use. In effect, the load line is drawn between the point of *maximum* collector *voltage* with *minimum* collector *current,* and *minimum* collector *voltage* with *maximum* collector *current.* Since you have a 15 volt battery for collector bias, you can get a maximum of 15 volts on the collector if 0 volts is dropped across the load, R_L. And this can only happen if there is zero collector current. One end of the load line then is at the point that shows 15 volts and 0 milliamperes. This is shown as point X on the curve (see page 4-78). The next extreme of the load line is found by determining the maximum current that can flow in the collector circuit.

the load line (cont.)

Since a 15-volt battery is being used, the maximum collector current that can flow is that which will cause a 15-volt drop across the load. Since the load resistor is 3300 ohms, this can be determined by Ohm's Law as I = 15 volts/3300 ohms, or 4.5 milliamperes. Now, since 4.5 milliamperes will drop 15 volts across the load, the voltage left at the collector will be zero. So, the other extreme of the load line will be plotted at 0 volts and 4.5 milliamperes, point Y. When points X and Y are connected, the load line is produced.

An operating point for the circuit and parts shown on page 4-77 can be chosen anywhere along this load line. Until now, operating points were chosen on the base current curves, but actually they can be between these curves as well. If that is done, however, you will have to approximate the base current, and the design will not be as accurate as if the point were chosen on a base current curve.

An operating point can be chosen anywhere along the load line

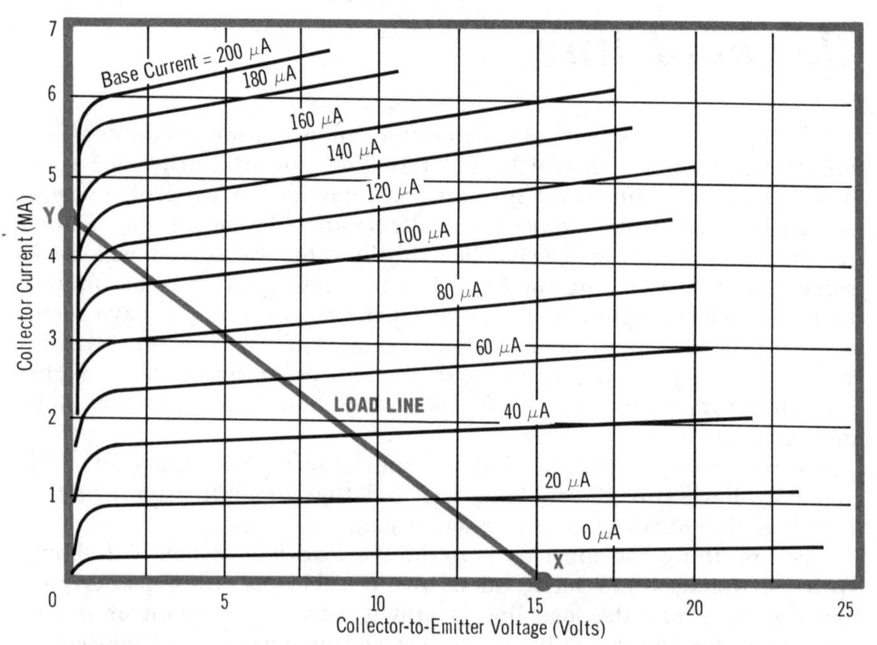

Although this load line was drawn based on a battery and load resistor on hand, you could first draw a load line, and then determine what bias battery and load resistor you would need. But, if you do it this way, you might come up with required battery and resistor values that are not readily available. It is best to start with standard parts, and compute the load line. If the parts you choose produce an undesirable load line, change the parts to shift the line to the desired position.

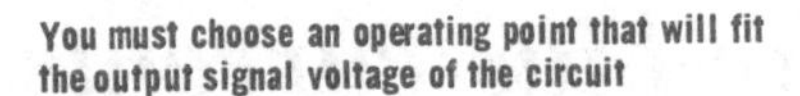

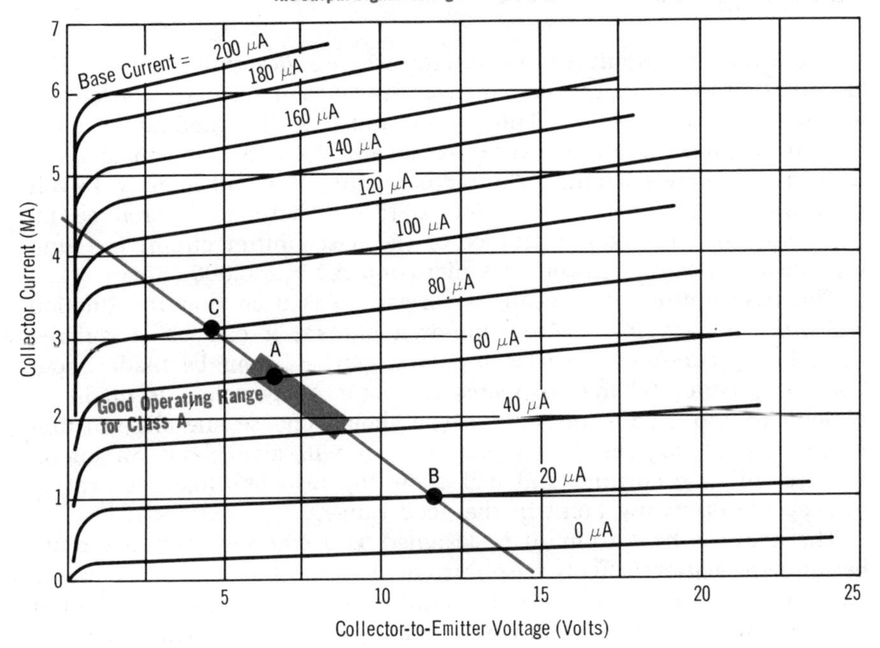

choosing an operating point

Some operating points might be better than others, depending on the circuit's purpose. For example, suppose the circuit is to be a class A amplifier, and the output signal voltage must swing 10 volts peak to peak. You would have to select the operating range shown to allow the collector voltage to vary between about 1 and 14 volts; this would permit a 5-volt swing on each side of the operating point, or 10 volts peak to peak, for any operating point in that range. The swing should never reach the actual extremes of the load line, or distortion will occur as the collector current reaches cutoff or saturation.

If you use point B as the operating point, which is at a collector voltage of about 12 volts, one half of a 10-volt peak-to-peak signal will only be able to go about 3 volts in one direction before it is clipped. Ordinarily, this would be considered distortion; but in some circuits, such as limiters or clippers, this is desirable, so you would want to use point B. (These circuits are covered in later volumes.) Of course, if you were just handling 10-volt pulses that go in only one direction, both points B and C would be good, depending on the pulse polarity. For convenience, let's choose the operating point at point A for 60 microamperes of base current. The center of the good operating range would be best, but then we would have to approximate the base current, which may not necessarily change linearly between curves.

setting the bias currents

The collector circuit of the circuit we are designing is complete because that was what the load line was drawn to match. The base circuit is not complete, however. This circuit must be designed to give the needed operating point. Suppose you choose point A. This means you must have a base current of 60 microamperes. Remember, only 15-volt batteries are available. Thus, for a 15-volt battery to produce 60 microamperes, the total resistance of the base-emitter circuit is found by Ohm's Law as R = 15 volts/60 microamperes = 250,000 ohms.

The base-emitter, you recall, is forward biased so that its junction resistance is very low (a few hundred ohms); it can be considered negligible. Therefore, the base input resistor, R_B, can be made about 250K to produce 60 microamperes for point A operation. It might be difficult to find a 250K resistor, so you should choose one close to that value. A 0.27-Meg resistor is fine since it will give about 55 microamperes of base current, and will allow the resistor's tolerance rating to keep the operating point in the good range.

The above is how a circuit is designed as a common-emitter circuit, because the battery affects base current directly. But, with a common-base circuit, it must be done differently: the battery controls emitter current. However, this is also done simply, when you remember that emitter current is the total of base and collector currents. The collector current at point A is about 2.5 milliamperes, so $I_E = I_C + I_B = 2.5$ milliamperes + 60 microamperes = 2.56 milliamperes. With Ohm's Law, the total emitter resistance should be: 15 volts/2.56 milliamperes, or 5869 ohms. With this low value of resistance, however, the resistance of the base-emitter junction is no longer insignificant. If the junction resistance is, say, 400 ohms, then the emitter resistor must be about 5400 ohms to give the proper emitter and base currents. (A 5600-ohm resistor is the practical equivalent of the 5400 ohms needed.)

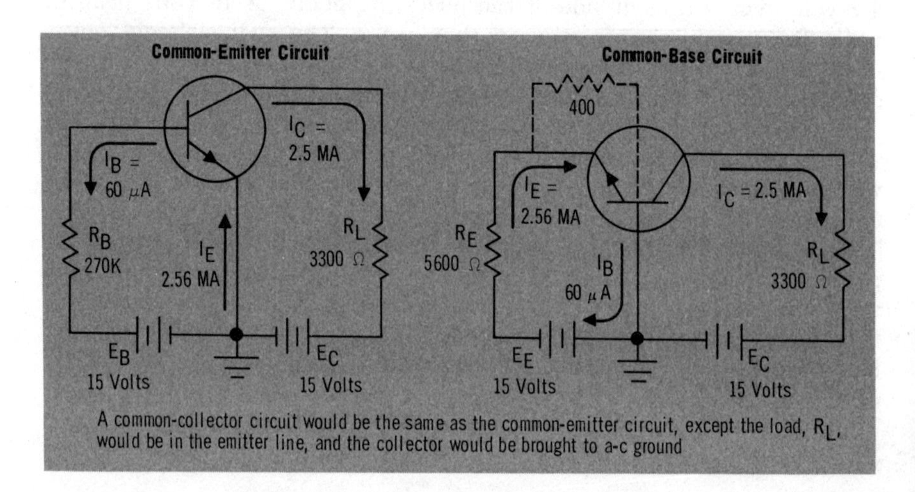

A common-collector circuit would be the same as the common-emitter circuit, except the load, R_L, would be in the emitter line, and the collector would be brought to a-c ground

changing values

For the example of designing the transistor circuit, the parts were, at first, arbitrarily chosen. Once the circuit is designed, the collector circuit cannot be changed without changing the load line. The base-emitter circuit, though, could be changed because it is just a specific bias current with which you are interested. For example, if you now wanted to use a 7.5-volt battery, the base resistor, R_B, could be changed to 133K to get a base current of 60 microamperes; and with a 3-volt battery, R_B would be 56K; and so on.

ALL THREE CIRCUITS WILL OPERATE ON POINT A OF THE LOAD LINE

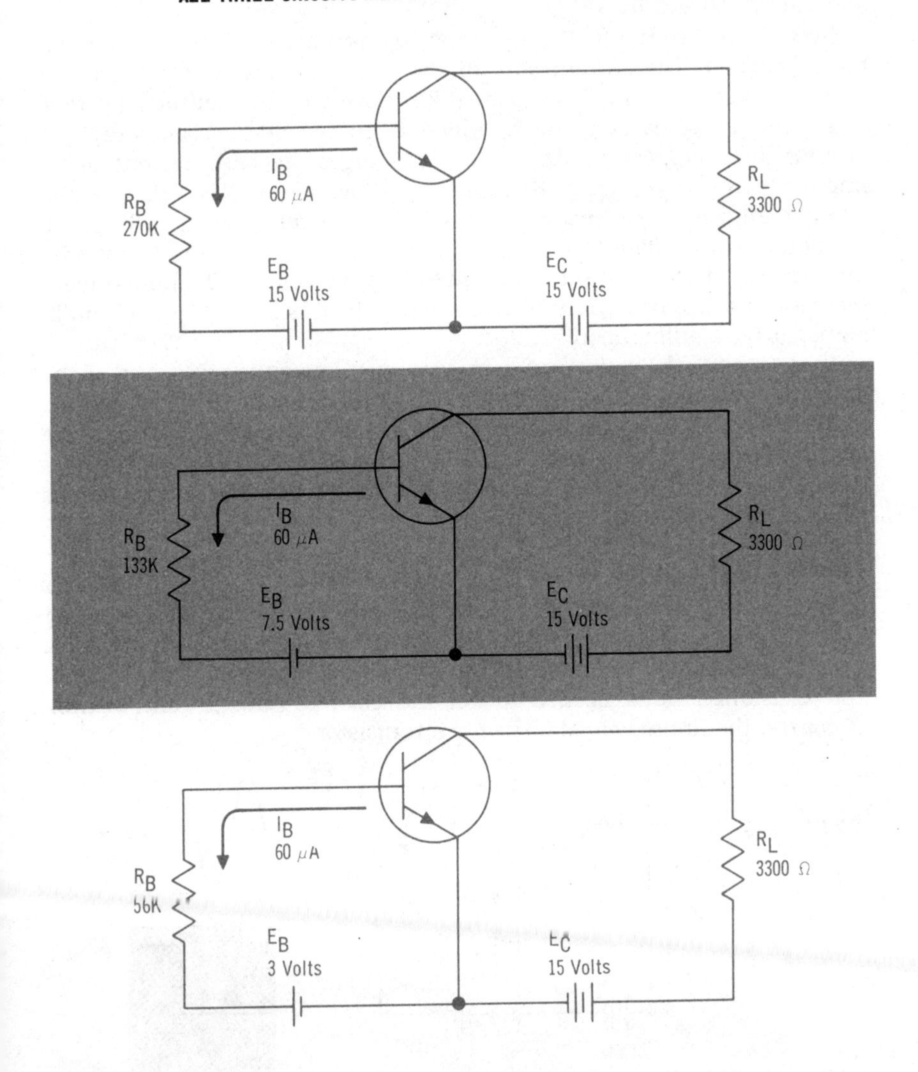

dynamic gain

Earlier, you learned that current gain is found by comparing output current to input current, and you used the actual d-c levels of current to find α and β. However, since you only used d-c, or static, levels of current, only the *static gains* were determined. With signal amplification, it is the *change* in output compared to the *change* in input that determines *dynamic gain*. A change in value is shown by the Greek letter delta (Δ), so a change in current is represented by ΔI, and a change in voltage is shown as ΔE. Actually, these represent the a-c signal current and voltage, which can be represented by another standard method: i and e. Lower case letters are used to represent ac, and capital letters for dc.

Now, let us compute the dynamic, or a-c, gains of the circuit we have designed. The current gain of the common-emitter circuit can be found by using the curves on page 4-79. Normally, the collector current is 2.5 milliamperes at point A, which is on the 60-microampere base current line. Suppose the input signal changes the base current to 80 microamperes. If you trace along the load line, you will find that on the 80-microampere base current curve, the collector current will go to 3 milliamperes. Therefore, for a 20-microampere (0.02-milliampere) increase in base current, the circuit produced a 0.5-milliampere increase in collector current. Current gain, then, is $\beta = i_C/i_B = 0.5$ milliamperes/0.02 milliamperes $= 25$.

To determine the resistance gain, you must know the resistance of the base-emitter junction, since this is the input resistance faced by the input signal. Earlier, we assumed this to be 400 ohms. As you were taught, the output resistance is the load, because that is what develops the output signal. Since the load is 3300 ohms, the resistance gain is $r_{GAIN} = 3300/400 \cong 8$.

Now the voltage gain can be found as $e_{GAIN} = \beta \times r_{GAIN} = 25 \times 8 = 200$. This means that a 20-millivolt input signal will cause the collector voltage to change 4 volts. If you wanted to increase the gain, another load line would have to be established with a larger collector voltage and load resistance. In this way, you could establish whatever current and resistance gains needed to get the desired voltage gain, within, of course, the limited capability of the transistor.

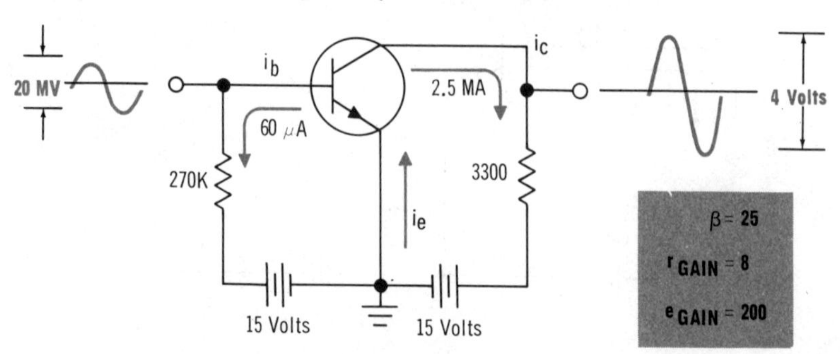

summary

☐ A transistor gives the desired gain only when it is used in a circuit with the proper parameters. ☐ Circuit parameters are indicated by a family of characteristic curves. ☐ Collector current, I_C, plotted against collector-to-emitter voltage, V_{CE}, for various base currents, I_B, is the most common family of characteristic curves. ☐ The maximum power dissipation curve indicates the safe current-handling capacity of the transistor under various operating conditions.

☐ Load lines show all possible operating points for a chosen collector voltage. ☐ The load line is drawn between the point of maximum collector voltage with minimum collector current, and minimum collector voltage with maximum collector current. ☐ The operating point is selected so that the input signal does not reach the extremes of the load line where the collector current is not linear. ☐ The load line sets the design of the collector-base circuit. ☐ The base current at the operating point sets the design of the base-emitter circuit in common-emitter circuits. ☐ The emitter current at the operating point sets the design of the base-emitter circuit in common-base circuits. ☐ The design of common-collector circuits is the same as that for common-emitter circuits except the load is in the emitter circuit and the collector is at a-c ground.

☐ Static gain is determined by d-c or static levels, and is represented by capital letters. ☐ Dynamic gain is determined by changes in the output for changes in the input. ☐ Dynamic gain uses a-c levels, and is represented by small letters. ☐ Varying the load line and the bias values changes the dynamic current and resistance gains. ☐ Dynamic voltage gain and circuit amplification are established by the dynamic current and resistance gains.

review questions

1. List the seven principal transistor design parameters.
2. What are *characteristic curves*?
3. Why are characteristic curves usually plotted around the base current?
4. Define the *maximum power dissipation curve.*
5. What does the load line show?
6. From what two points is the load line drawn?
7. What is the *operating point*? How is it selected?
8. Why is the maximum power dissipation curve needed?
9. How does *dynamic gain* differ from *static gain*?
10. How is a set transistor voltage gain obtained?

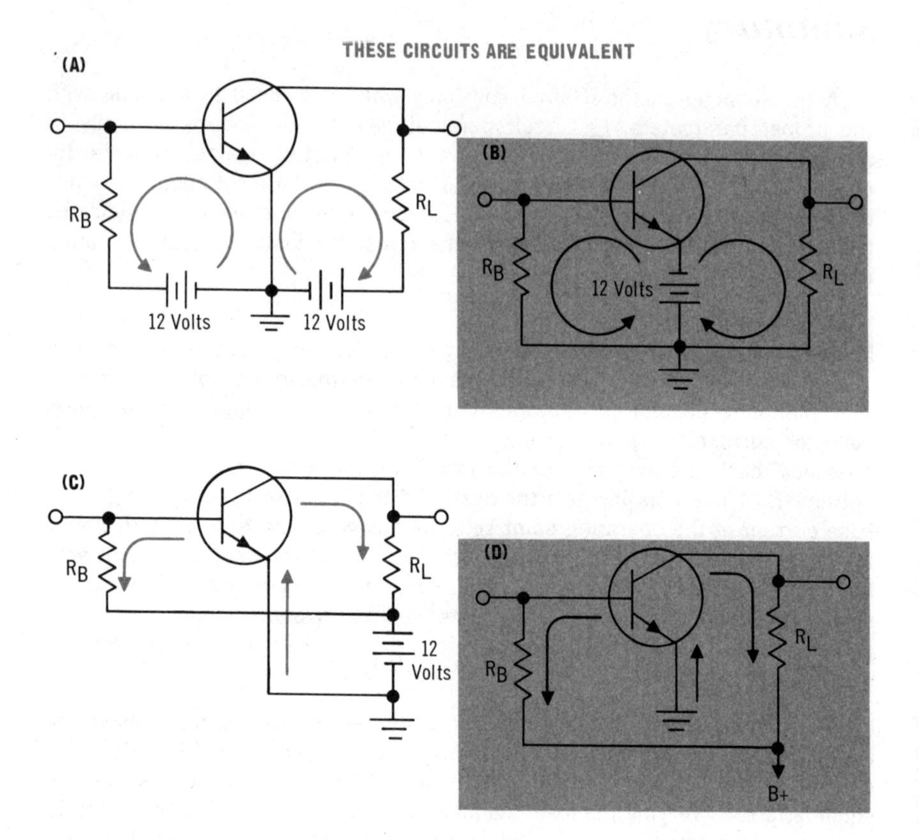

using one bias battery

Until now, the input and output circuits of the transistor used their own battery to produce the bias currents. The value of the collector battery is important in establishing a load line, but, as you learned, the actual voltage of the base-emitter battery is not that important. The circuit on page 4-82 shows that the emitter has the same polarity with respect to the base as it does to the collector. This permits the use of only one battery to establish the same circuit parameters as with two batteries. The circuit operation is essentially the same, except one bias battery supplies two paths, and the values of the resistances determine the parameters. The size of the battery establishes the maximum collector voltage for the load line, and the value of R_B sets the base current. As the diagrams show, this can be done in two ways. Circuit C is the one most often used. Circuit D shows how circuit C is usually drawn with a regular supply voltage on overall schematic diagrams, where a common remote power supply is used to feed a number of different circuits.

fixed bias

In all of the circuits you have been studying thus far, the bias currents for the input and output circuits were derived directly from the battery; the input circuit was in series with a resistor to limit the input bias current to the desired level. This type of bias is known as *fixed bias,* since a fixed d-c voltage source and resistance path are used.

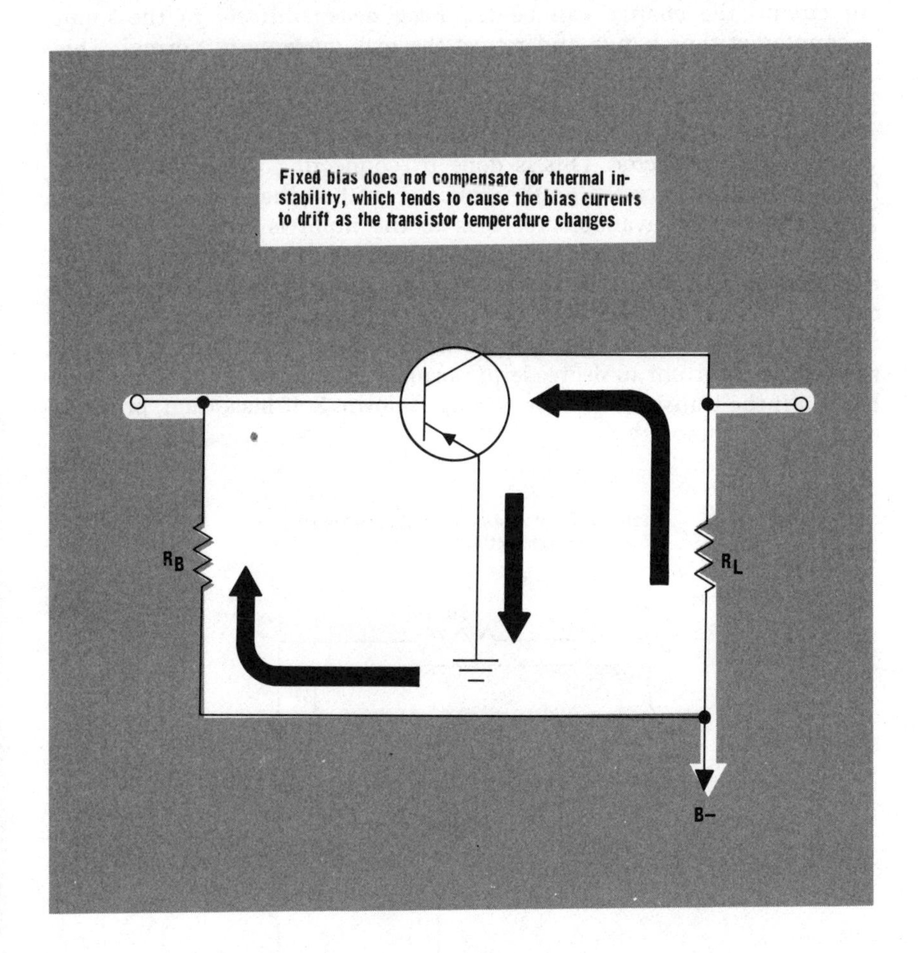

Fixed bias has some drawbacks because it does not compensate for variations in the static bias currents that may occur due to changing circuit characteristics. For example, as the temperature of the transistor changes, its junction resistances will also change, varying the bias currents. Any change in bias current will shift the operating point of the circuit, which, in turn, can adversely affect the gain of the transistor. This is known as *thermal instability,* and fixed bias is greatly affected by it.

self bias

The ability of a circuit to compensate for temperature changes is known as *thermal stability*. The transistor is naturally affected by temperature, and so to provide for thermal stability, any unwanted change in bias current that takes place should be used to *counteract* the change. Since any change in bias current will show up in the output circuit, the change can be *fed back degeneratively* to the input to counteract the change and return the bias currents to normal. This is also known as *negative feedback*.

One way to do this is with *self bias*. This is so called because the actual bias current in the input circuit is determined primarily by the voltage at the collector. This is done by connecting the input current-limiting resistor directly to the collector rather than to the battery. Then, the voltage available for bias to the input is what is left over after the drop occurs across the output load. So, if temperature causes the transistor bias current to rise, the increased collector current will cause a bigger drop across the output load, and the voltage at the collector will go down. This will reduce the input bias current, causing the collector current to decrease to normal. Of course, the opposite will happen if the transistor temperature goes down. Self bias, then, provides some thermal stability.

Self bias is obtained when the input bias is gotten from the collector instead of the battery

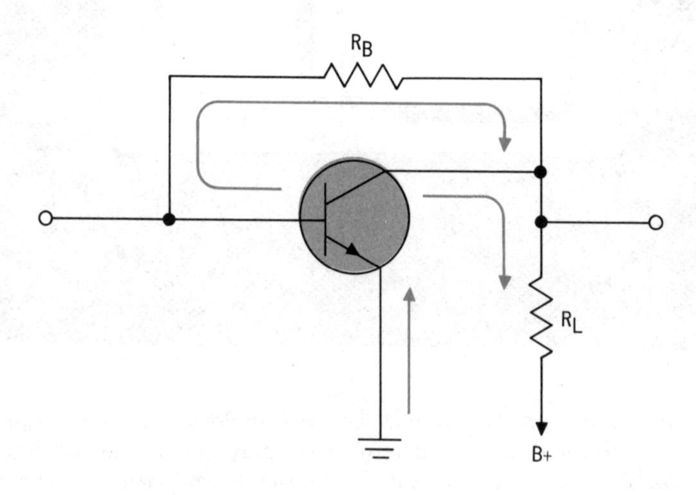

Changes in collector voltage, therefore, will be fed back to the input to stabilize the transistor against temperature variations

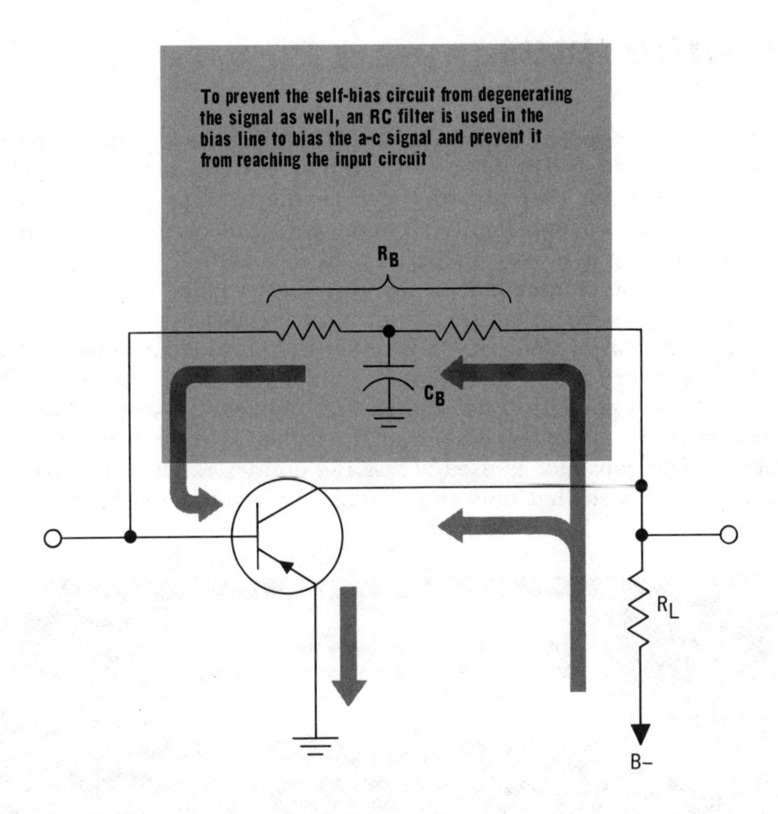

degenerative feedback

Self bias is better than fixed bias because it uses negative feedback to bias itself and thus compensate for thermal changes. However, self bias, as shown on the previous page, also has a disadvantage. Since it uses degenerative feedback, the amplified *signal* at the collector is also fed back to reduce the input signal. This essentially reduces the gain of the stage. The self-bias circuit, then, should be arranged to feed back back only the d-c changes at the collector, and not the a-c changes. This is simply accomplished by using an RC filter in place of the bias current-limiting resistor. The capacitor in the filter will bypass any a-c signal to ground so that it will not degenerate the input signal. The gain of the stage will not be affected by the feedback.

The d-c feedback, still exists though, since the d-c current through R_B is still controlled by the static, or average, level of collector voltage. A centertapped resistor, or two resistors, must be used to isolate the capacitor from the input and output circuits. Otherwise, the capacitor would bypass the input or output circuits. As shown, the capacitor only bypasses the feedback line.

emitter bias

Another method of obtaining a bias voltage to produce thermal stability is *emitter bias*. Since the emitter current is the total of the base and collector currents, any drift in the average d-c level of bias current in the common-emitter circuit will also occur in the emitter line. By putting a resistor in series with the emitter, a voltage will be produced at the emitter that opposes the bias voltage at the base. The values of R_B and R_E in such a circuit must be chosen so that the proper base-emitter bias current will flow under ordinary circumstances. Then, if the transistor temperature changes to increase bias currents, the drop across R_E will go up to raise the emitter voltage. This will oppose the input bias to reduce the base, and thus, the collector bias current to normal. The capacitor is used across the emitter resistor to bypass the signal voltages so that only the average d-c level is fed back and the gain of the stage is not reduced.

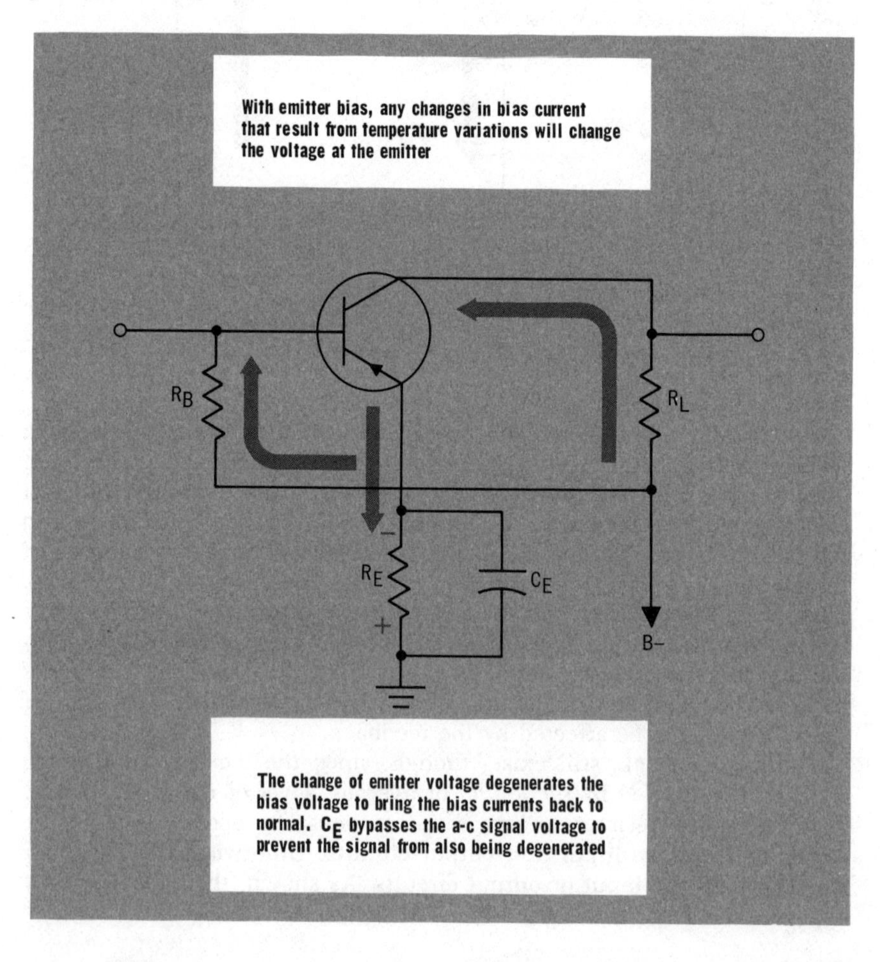

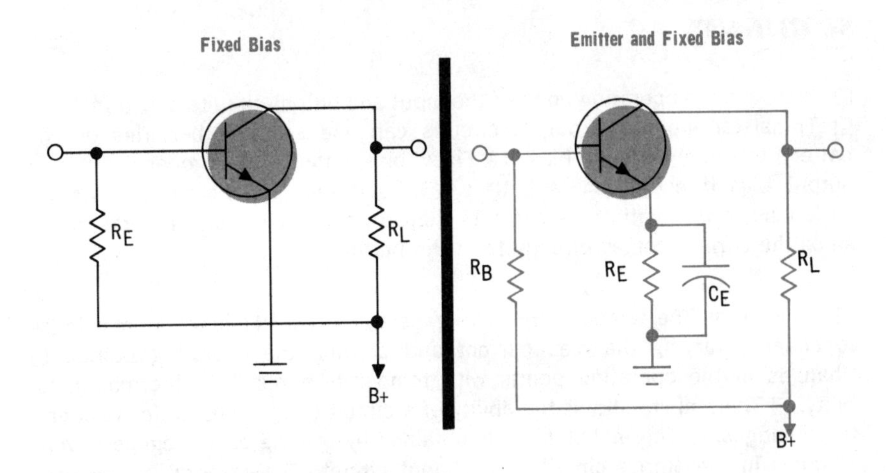

combination bias

Actually, as noted on the previous page, emitter bias was not used by itself; it was combined with fixed bias. The fixed bias was used to set the base-emitter current, and the emitter bias merely modifies that current for stability. In this way, fixed bias can be used to get some degree of thermal stability. In the same way, emitter bias can be combined with self bias to improve the thermal stability of the circuit even further.

Emitter bias cannot be used alone because it really does not provide an input current path; it only produces a degenerative voltage. Self bias and fixed bias could be combined, but nothing would be gained by this because, for self bias to be effective, it would have to be the major path for the bias current, and the fixed bias resistor would only serve as an unnecessary voltage divider.

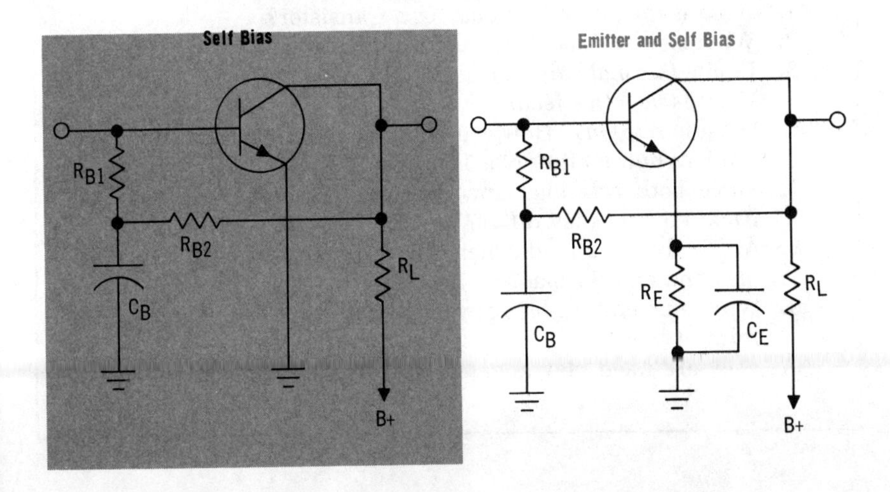

summary

☐ Bias sets the operating point of the input and output circuits of a transistor. ☐ Transistor input and output circuits can use separate batteries or one battery to supply proper bias. ☐ Fixed bias obtains the proper input and output bias directly from a battery. ☐ Fixed bias cannot compensate for variations in the static bias currents due to changing circuit characteristics, since the current comes directly from the battery.

☐ Changes in the temperature of a transistor cause its junction resistance to change, varying the bias current and shifting the operating points. ☐ Changes in the operating points with temperature result in thermal instability. ☐ Thermal stability is the ability of a circuit to compensate for temperature changes. ☐ Thermal stability is obtained by feeding back, degeneratively, changes in the output circuit to the input circuit. ☐ Degenerative feedback is also known as negative feedback. ☐ Self bias obtains the proper input bias with negative feedback from the collector circuit. Self bias feeds back unwanted changes in the collector voltage to the base to bring the input current back to normal.

☐ Thermal stability can also be obtained with emitter bias. ☐ Emitter bias feeds back undesirable changes in the emitter voltage, due to changes in the input current, to the base to bring the input current back to normal. ☐ Emitter bias must always be combined with another form of bias to provide an input current path. ☐ Emitter bias is combined with fixed bias to provide thermal stability, or it can be combined with self bias to improve thermal stability.

review questions

1. What is the purpose of bias for a transistor?
2. What is *fixed bias*? What is its main disadvantage?
3. Define *thermal stability*.
4. What is *negative feedback*?
5. What is *self bias*? How is it obtained in a transistor circuit?
6. Explain how emitter bias is produced.
7. Since both self bias and emitter bias use negative feedback, how do they differ?
8. Are both a-c and d-c negative feedback supplied by self bias and emitter bias?
9. Why must emitter bias always be combined with some other form of bias?
10. Why are fixed bias and self bias usually not combined?

stabilization

The bias circuits that provide thermal stability compensate for the effect of temperature by responding to changes in current *after* they occur. They do so because the changes must occur first, to some degree, before they can be fed back for compensation. Also, because of this, complete compensation is impossible because, if the static bias currents were actually returned to their proper levels, the feedback would be lost. Actually, then, the bias circuits only oppose the changes due to thermal instability in an attempt to keep the variations that occur within tolerable limits.

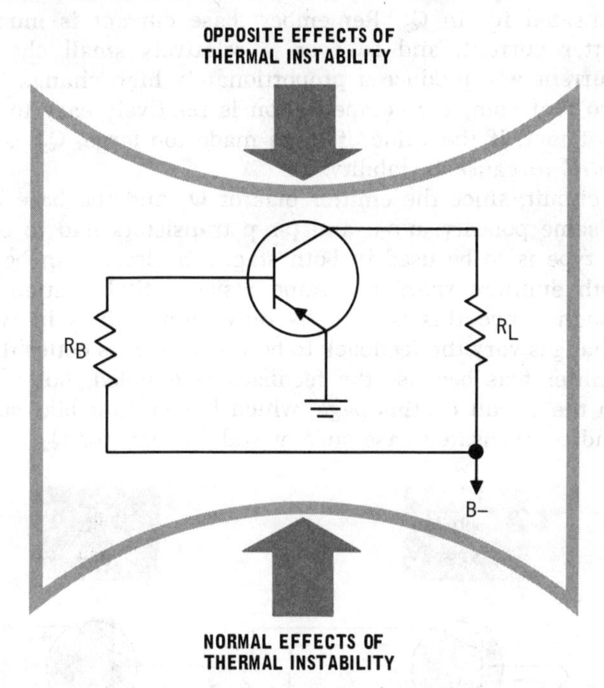

A good stabilizing circuit will produce effects that are opposite to those of ordinary temperature changes, and so will cancel thermal instability before it occurs

Therefore, stabilizing circuits, to be more effective, should not rely on feedback that results from thermal instability; they should prevent the bias currents from changing in the first place. One way to do this is to produce an effect on the transistor that is opposite to the normal effect of thermal instability. The following are ways this can be accomplished:

1. Use one transistor to stabilize another.
2. Use a thermistor to control bias current.
3. Use a diode to control emitter-base current.
4. Use a diode to control collector-base current.
5. Use combinations of stabilizing circuits.

common circuit stabilization

One way to accomplish better stabilization is to have the individual transistor stages share a *common circuit*, as shown. These circuits use common bias paths for both stages. In the circuit, the B-minus supply and resistor R_E provide bias currents for the emitter of Q_1 and the base of Q_2. If the temperature goes up to increase the emitter current of Q_1, the drop across R_E will also go up, and thereby leave less voltage to be applied to the base of Q_2. The base bias current in Q_2, then, will go down. The effect of temperature on Q_1, as a result, was produced in an opposite way in Q_2, so that any gain changes that take place in Q_1 will be compensated for in Q_2. Remember, base current is much smaller than emitter current, and so even a relatively small change in Q_1 emitter current will produce a proportionately high change in Q_2 base current, so that complete compensation is relatively easy to attain. As a matter of fact, if the value of R_E is made too large, Q_2 can be *over-compensated* to cause instability.

In the circuit, since the emitter bias of Q_1 and the base bias of Q_2 have the same polarity, n-p-n and p-n-p transistors had to be used. If the same type is to be used in both stages, a circuit can be employed where both emitters share the same resistor. Stabilization is not as good, though, since this is actually only emitter bias in which both current changes vary the feedback to both stages. It is better than single-circuit emitter bias because the feedback is doubled, but not as good as that in the circuit on this page, which has emitter bias stabilization for Q_1, and compensated base current stabilization for Q_2.

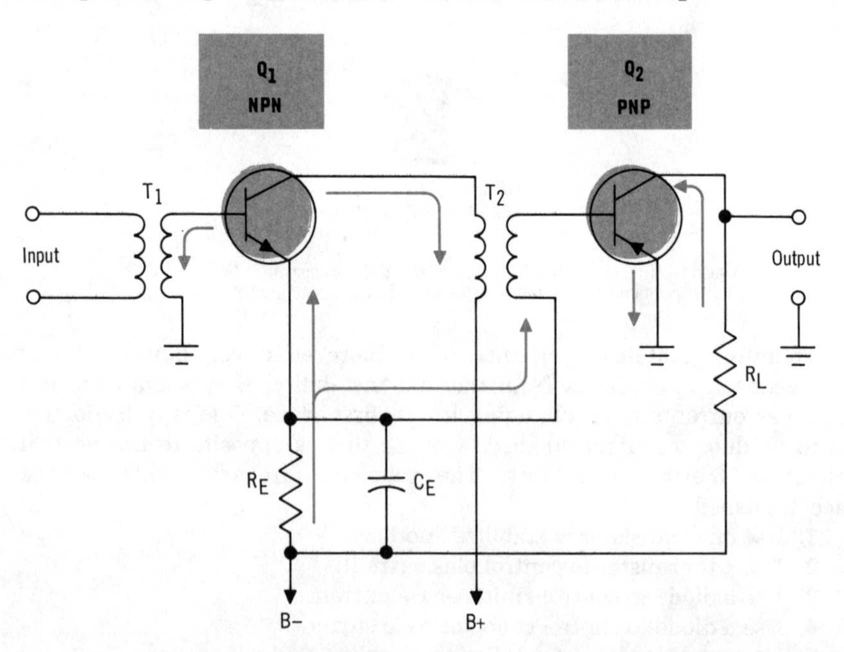

thermistor stabilization

It is possible to use one transistor to stabilize another because they are both similarly affected by temperature. Also, any other part could be used to stabilize a transistor if that part too were affected by temperature similar to a transistor. One such device is the *thermistor*, which, as you learned earlier, also has its resistance reduced as temperature rises. One use of the thermistor is shown in A. The thermistor forms a voltage divider with R_1 to provide fixed bias for the base. Then, as temperature goes up, tending to increase base bias current, the resistance of the thermistor goes down. Thus, it drops less voltage for the base, so that the base bias current tends to go down. When the circuit is properly designed, the thermistor tends to reduce the base current by the same amount that the transistor temperature tends to raise it; so, base bias current stays constant.

In B, the thermistor is used as part of a voltage divider with an emitter bias resistor. The base is supplied with fixed bias, and the emitter bias provides the usual stabilization. But the action of the thermistor increases the emitter stabilization for complete compensation. When the temperature goes up, tending to increase base current, the resistance of the thermistor goes down, causing the emitter to go more negative. This reduces the base-emitter bias sufficiently, so that the base current is lowered by the same amount it tended to increase.

In both circuits, since the thermistor and the transistor are affected by temperature at the same time, compensation takes place before the bias currents can change.

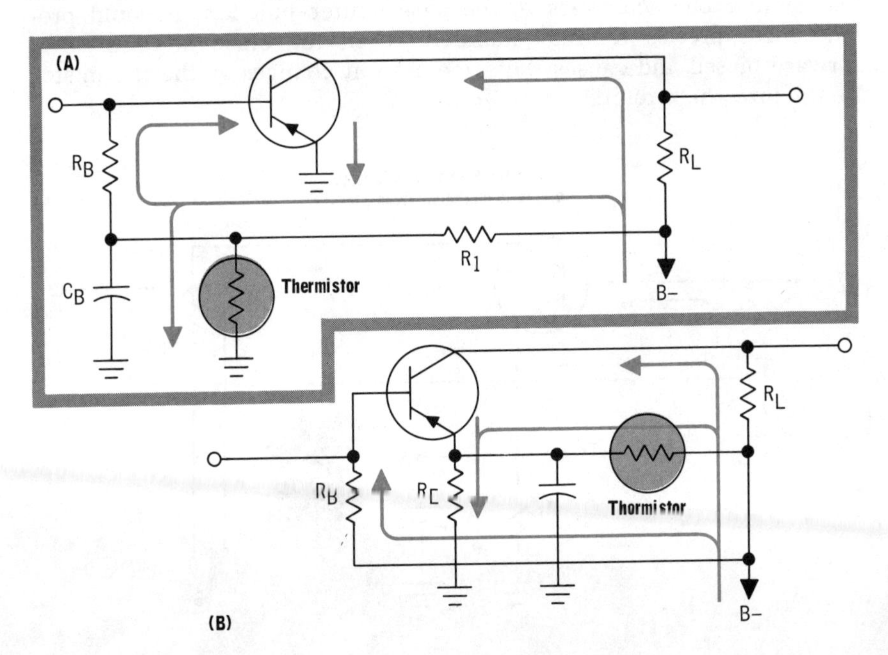

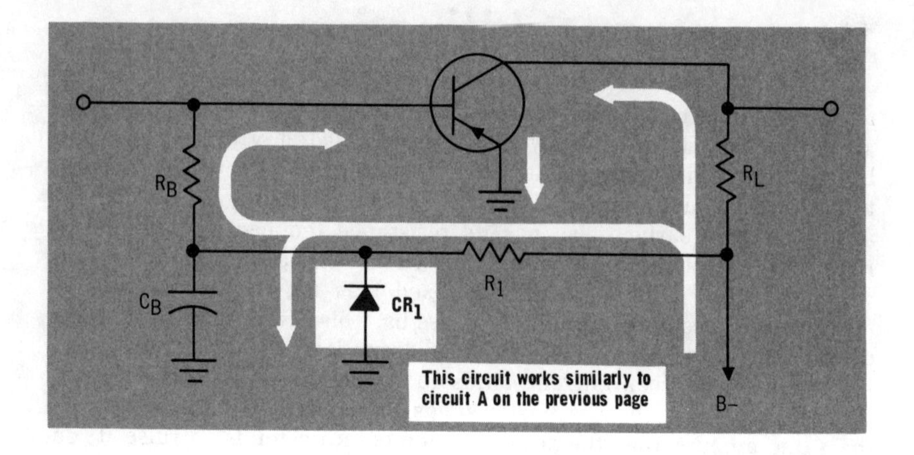

diode stabilization

In the previous circuit, the best results will be obtained when the *negative temperature coefficient* of the thermistor most nearly matches that of the transistor. However, it is very difficult to have a thermistor act exactly like a transistor over a range of temperature variations since both devices are not near enough alike physically. The emitter-base circuit of the transistor that is being compensated is actually a p-n junction, and the thermistor is just a slab of semiconductor material. So, if a p-n diode were used in place of the thermistor, and the diode had similar characteristics to the base-emitter junction, it could provide more precise thermal stability. When the diode is used, it is forward biased and causes the same current changes as the thermistor to stabilize the circuit.

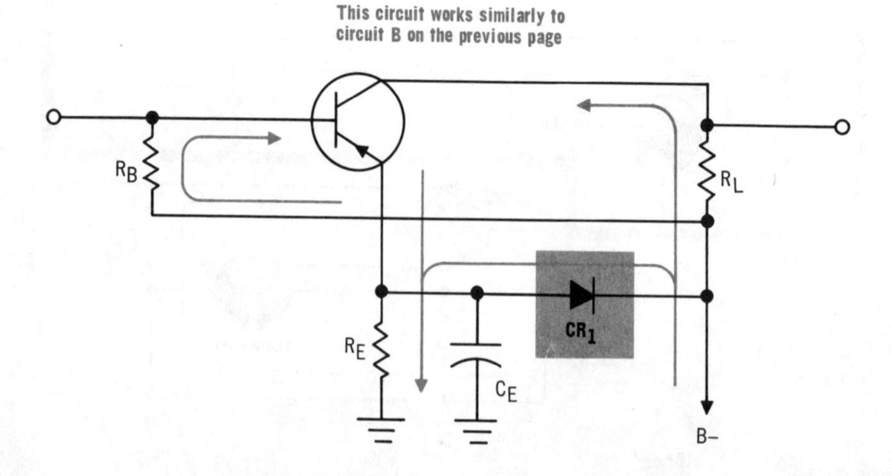

collector-base current stability

The previous diode stabilizing circuit only compensated for changes in base-emitter bias current, and thereby prevented the collector current from varying with such changes. However, the collector current itself can change with temperature, although the base-emitter current is stabilized. This is because the previous diode stabilizing circuit is a forward biased circuit, and so only compensated for changes in forward, or majority, current. But, as you learned in semiconductor theory, when temperature goes up, it also increases *minority carriers*. The minority carriers are increased in both the base and collector, but since the actual collector current is determined by the number of minority carriers in the base, those are the ones for which compensation is needed.

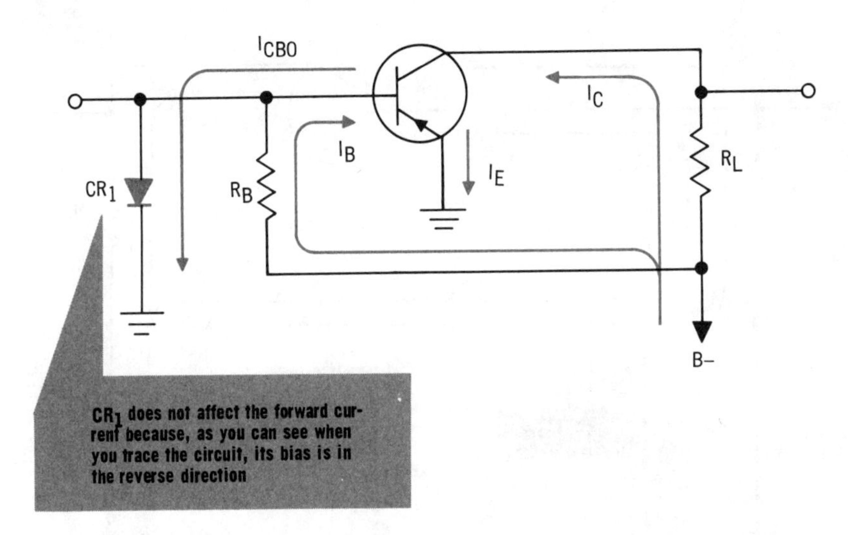

Actually, there are two currents flowing in the base: collector-base current in one direction, and emitter-base current in the other direction. The difference between the two, which is known as the base current, I_B, is a very small current that flows in the same direction as emitter-base current, since that is the larger of the two. In any event, if the base-collector current, which is known as I_{CBO}, changes with temperature, it will cause I_B to change also, but in the opposite direction because I_B is a *difference* current. To compensate for this, a reverse biased diode is connected in the base circuit to match the temperature characteristics of the reverse biased base-collector junction. When temperature goes up, and the number of minority carriers are increased in the base, the resistance of the diode goes down to drain off the excess minority carrier current (I_{CBO}), so that the difference, or base current (I_B), remains the same.

combined diode stabilization

You can see that for complete thermal stability, both I_B and I_{CBO} must be compensated for when diodes are used. This is done by using two diodes in the base bias circuit. One diode is forward biased to control base-emitter current, and the other diode is reverse biased to control base-collector current. The resistances of both diodes go down with temperature. But, since diode CR_1 is forward biased, it has a much lower resistance than diode CR_2; so, CR_1 has much more effect in the voltage divider line. Diode CR_1 lowers the voltage when the temperature goes up to keep the base-emitter current from rising with the temperature. Since diode CR_1 is forward biased, it conducts forward, or majority carrier, current.

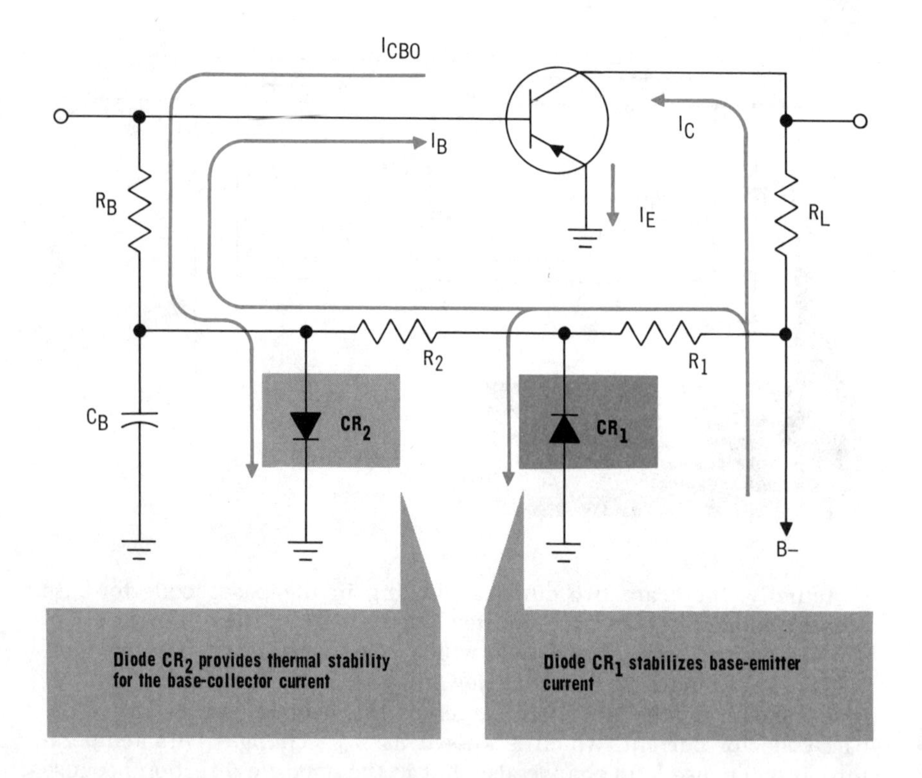

Diode CR_2 is reverse biased, has a high resistance, and conducts reverse, or minority carrier, current; so it does not affect the majority carriers of the base-emitter circuit. It does, though, pass the base-collector minority current, and drains off the extra minority carriers produced in the base when temperature rises; thus it keeps I_{CBO} constant.

Both diodes are needed for the circuit to have full thermal stability.

summary

☐ Simple bias stabilization circuits provide thermal stability by responding to changes in output current after they occur. ☐ Bias stabilization circuits can use semiconductors to prevent the current from changing in the first place. ☐ Semiconductors cancel the effect of thermal instability by producing an effect opposite to that of thermal instability. ☐ One transistor can stabilize another through a common bias circuit. ☐ The bias is generated by the compensating transistor and is applied to the compensated transistor to provide either emitter bias or base current stabilization. ☐ Base current stabilization provides complete compensation; emitter bias stabilization only provides increased bias feedback.

☐ Thermistor semiconductor devices can also stabilize a transistor since both are similarly affected by temperature variations. ☐ The thermistor replaces a resistor to supply fixed bias to the base that decreases with increasing temperatures or to supply reverse emitter bias that increases with increasing temperatures.

☐ P-n diodes can stabilize a transistor since they possess a p-n junction that is similarly affected by temperature variations. ☐ A forward-biased diode replaces a thermistor to provide fixed or emitter bias that varies with changes in temperature. ☐ Full thermal stability of a transistor is achieved by compensating both the emitter-base and the collector-base junctions. ☐ The collector-base junction is compensated for increases in minority carriers due to temperature variations. ☐ A reverse-biased diode is connected to the base circuit to drain off any excess collector minority carriers caused by increasing temperatures.

review questions

1. What is the main disadvantage of simple bias stabilization circuits?
2. How do semiconductors provide bias stabilization?
3. List 5 types of semiconductor stabilization circuits.
4. What is *common circuit stabilization*?
5. Why is base current stabilization superior to emitter bias stabilization?
6. How is a transistor overcompensated?
7. What is *thermistor stabilization*?
8. How does a diode compensate an emitter-base junction?
9. How does a diode compensate for thermal instability of minority carriers?
10. Why are two diodes always needed for full thermal stabilization?

circuit configurations

The basic transistor circuits you have studied thus far were built up slowly and methodically, so that by now you should be able to recognize the types of circuits and the kinds of bias used. However, when you use many commercial schematic diagrams, the circuits may not be as easy to recognize because they can be drawn in so many different configurations.

It would be good practice to redraw some commercial circuits to make them look like the basic circuits you have studied. This will help you to recognize various circuit configurations, and even help you to understand more about circuit theory.

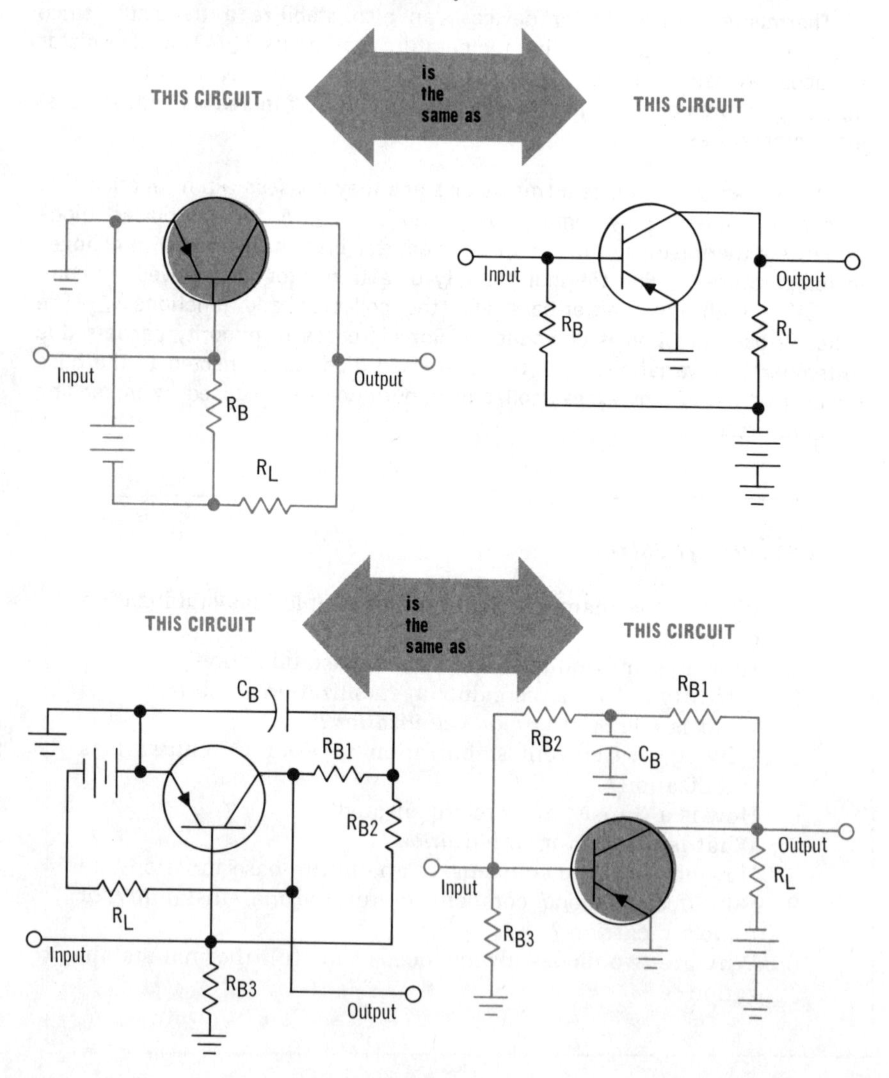

circuit configurations (cont.)

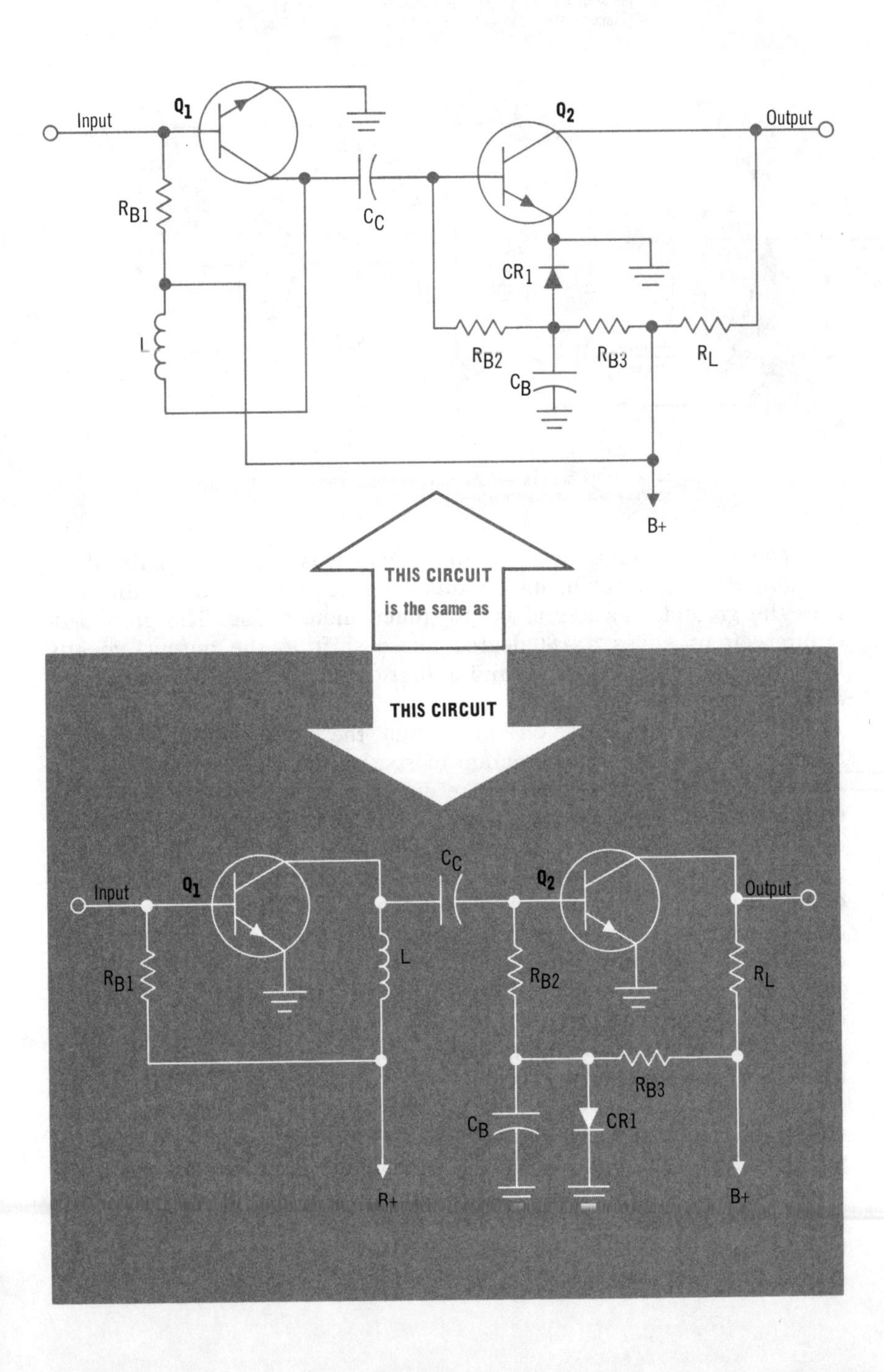

oscillation

Since the input and output signals of a grounded base circuit are in phase, that circuit is not as stable for amplifying, but is much more useful for oscillating

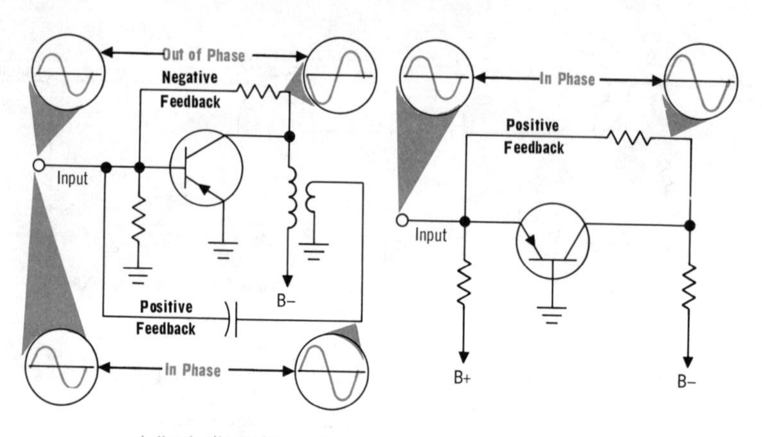

In the circuits, positive feedback means regenerative feedback, and negative feedback means degenerative feedback

A common-, or grounded-, emitter circuit was used to explain all of the amplifier and stabilizing circuits because it has better gain than does the grounded-base, and is also much more stable. The grounded-emitter circuit gives a 180-degree phase shift to the output, so any interaction between the input and output circuits will only cause degeneration.

With the grounded-base circeuit, though, the input and output signals are in phase, so if any interaction exists, *regeneration* will take place. For example, if in a grounded-base circuit, the input signal raises the emitter current, this will increase the collector current. Then because of the bias polarity, collector voltage will go in a positive direction. If part of this positive swinging output voltage is fed back to the emitter circuit, it will raise the emitter current again, which in turn wll increase the collector current, and so on. The output then is said to reinforce or regenerate the input, and the current will run away with itself and cause the circuit to generate its own signals. This is known as *oscillation,* and is more completely taught in a later volume.

The tendency to oscillate prevents us from using the feedback stabilizing circuits with the grounded-base circuit that we can use with the grounded-emitter circuits. However, this is not a completely undesirable characteristic. This is why the grounded-base circuit is most popular with *oscillator* circuits. The grounded-emitter circuit can be made to oscillate, but the output signal, which is 180-degrees out of phase with the input signal, must first be inverted. This could be done with a transformer or another transistor; but with a common-base circuit, it can be done simply by adding a self-bias circuit.

classes of operation

You learned in Volume 3 that electron tubes have certain classes of operation that can be used to obtain certain characteristics. Transistors can be operated in the same ways to get the same results. The emitter voltage – collector current curve of a transistor shows how the transistor can be operated class A for linearity, class B for rectification, and class C for efficiency. The circuits that employ these specific applications are covered in later volumes. The classes of operation for a transistor do differ somewhat from those of a tube, though, for two reasons. One is the reverse-current characteristic, and the other is its forward-current saturation operation.

Class C operation cannot be too far from the cutoff point because then the size of the input signal that would be required to drive the circuit into conduction would also drive the emitter circuit into the zener breakdown region. The bias point and the amplitude of the input signal are therefore limited by the reverse-current characteristics of the input circuit.

Transistors have more useful saturation characteristics than tubes because they are less prone to damage from saturation current, and the impedance of the output circuit can drop to about a few hundred ohms through collector current saturation. This makes the transistor useful in pulse switching circuits.

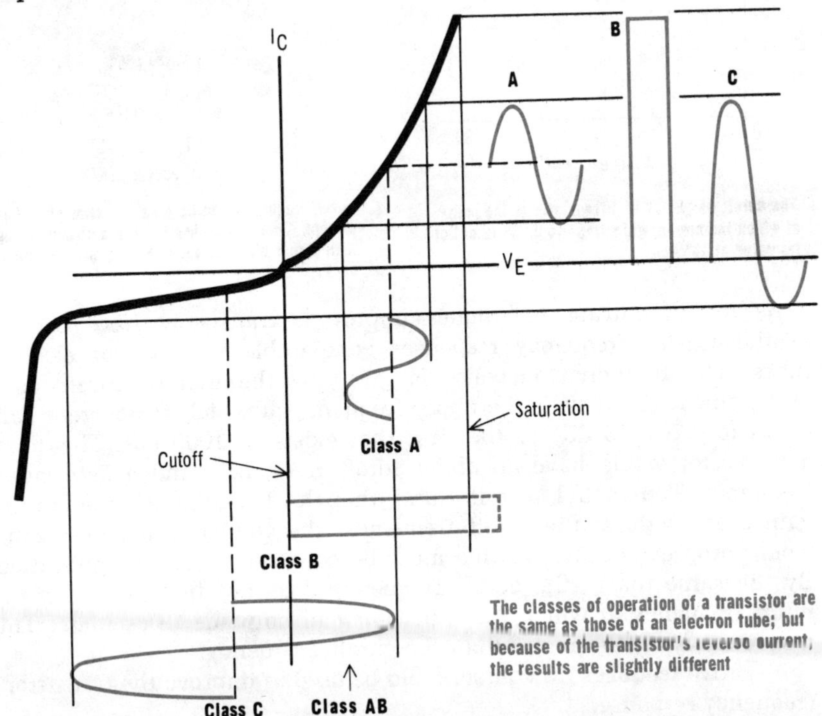

The classes of operation of a transistor are the same as those of an electron tube; but because of the transistor's reverse current, the results are slightly different

frequency response

One of the early major disadvantages of the transistor was its poor frequency response. The frequencies of the signal that the transistor can handle are determined by the transit time of the current carriers from the emitter through the collector. This is analogous to the transit time of the electrons from the cathode to the plate in the electron tube. The faster the current carriers move through the transistor, the higher the frequency they can handle. Since the velocity, or *mobility,* of the carriers depends, to a large extent, on the attractive forces of the bias voltages, better frequency responses can be obtained with higher voltages. In addition, since free electrons are easier to free and move than holes, n-p-n transistors have better frequency responses than comparable p-n-p transistors.

Another characteristic of the transistor that affects frequency response is the capacitance across each of the junctions, which tends to bypass higher frequency signals to ground. These *junction capacitances* are kept low to improve frequency response by making the base as thin as possible.

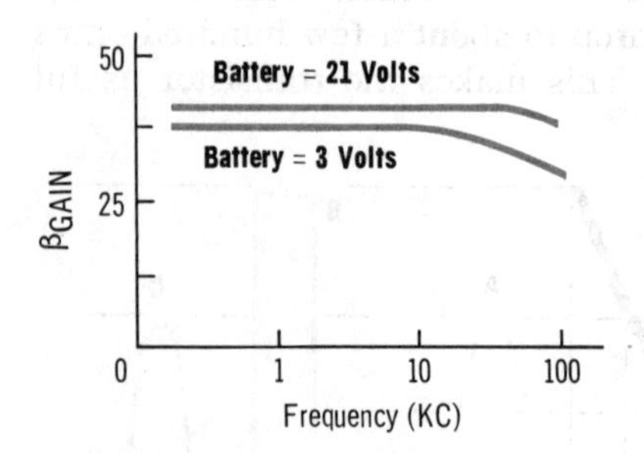

The cutoff frequency of a transistor is the point at which the current gain drops to 0.707 (3 db) of its value at 1000 cps

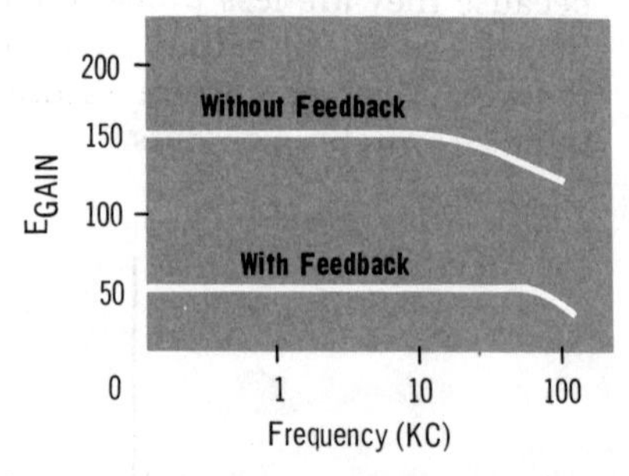

Cutoff frequency can be given in terms of alpha or beta. This can be improved in a circuit, though, with higher bias voltages and degenerative feedback

As manufacturing techniques improved, transistors have become available with frequency responses comparable to those of electron tubes. The frequency response is given by the manufacturers as a *cutoff frequency.* This is the upper frequency at which the current gain drops to 0.707 (3 db) of the gain that exists at 1000 cps. Therefore, a transistor would have an *alpha* cutoff frequency, and a *beta* cutoff frequency. You should keep in mind, though, that although the current gain drops 3 db at the cutoff frequency, the voltage and power gains would drop even more, so this must be considered in any application. By the same token, the cutoff frequency does *not* indicate the maximum frequency at which the transistor can be made to oscillate. This frequency is much greater than the cutoff frequency.

Negative feedback, of course, could be used to improve the transistor's frequency response.

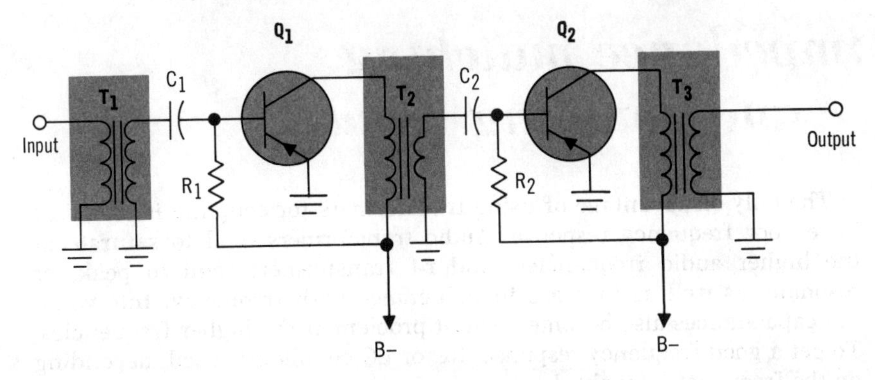

impedance matching and coupling

A transistor circuit has a low input impedance, up to around 1000 ohms, and a relatively high output impedance, about 20,000 ohms. These impedances must be considered when transistor circuits are used in *cascade,* since to get the maximum transfer of signal strengths, the output impedance of one stage should match the input impedance of the next. This *impedance match* is not always easy since the impedances are not close together. If the stages are connected directly, the input impedance of the second stage will shunt and lower the output impedance of the first stage and reduce its gain.

Probably the easiest way to match impedances is by use of transformers, with the primary winding acting as the high-output impedance of the first stage, and the secondary winding matching the low-input impedance of the first stage. Although this would require a step-down voltage action, it provides a step-up *current gain;* and since the transistor is current-sensitive, the overall gain is improved with *transformer coupling.*

You will probably notice that many transistor circuits use tapped transformers for loads. This is done so that the transformers can be designed for good coupling and gain purposes, and the taps can be provided for the best impedance matching.

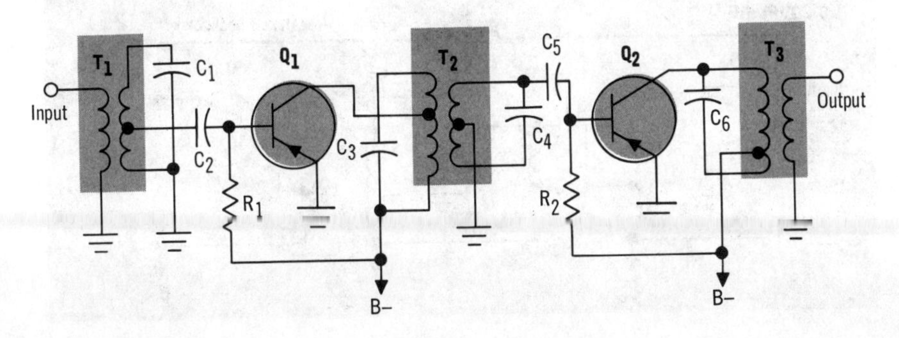

impedance matching and coupling (cont.)

The only disadvantage of using transformers for coupling is that they have poor frequency response. Audio transformers tend to saturate at the higher audio frequencies, and r-f transformers tend to peak, or resonate, as well as increase in inductance with frequency. Interwinding capacitances also become a shunt problem at the higher frequencies. To get a good frequency response, RC or LC coupling is used, depending on the frequencies involved.

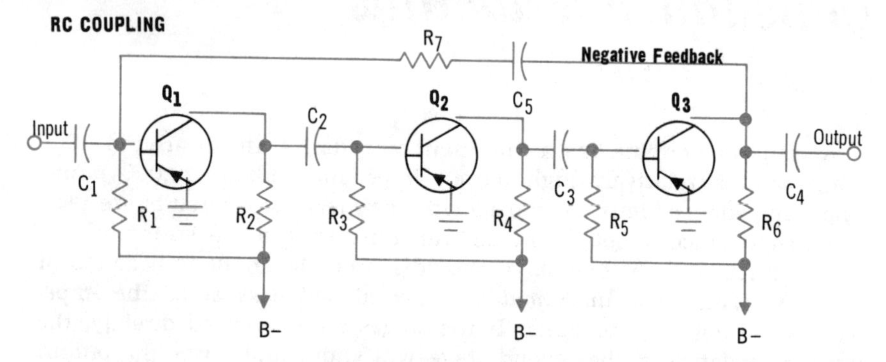

When *RC* or *LC* coupling is used, impedance matching must be compromised for frequency response. As a result, the gains of the stages are somewhat reduced. Because of this, the stages must be designed for greater gain in the first place to compensate for the loss, or an extra stage must be added to give the extra gain. With LC coupling, though, tapped coils can be used to obtain the proper impedances through *autotransformer* action. The coupling networks can be designed for better impedance matching also if negative feedback is used to broaden the frequency response. But this, too, lowers the gain, and might require an additional stage.

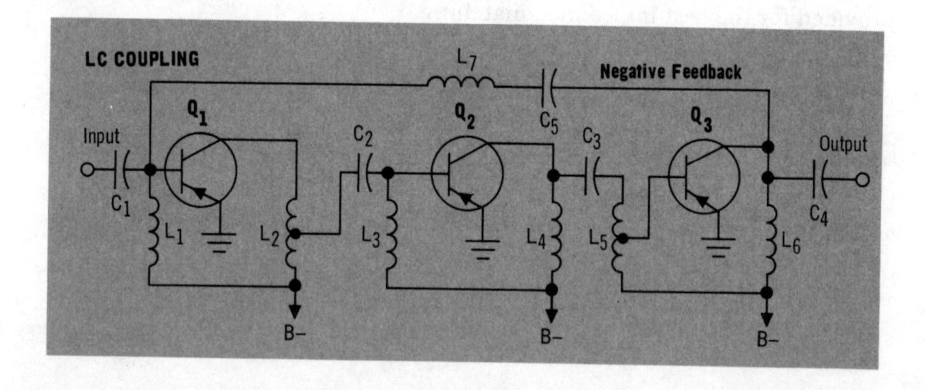

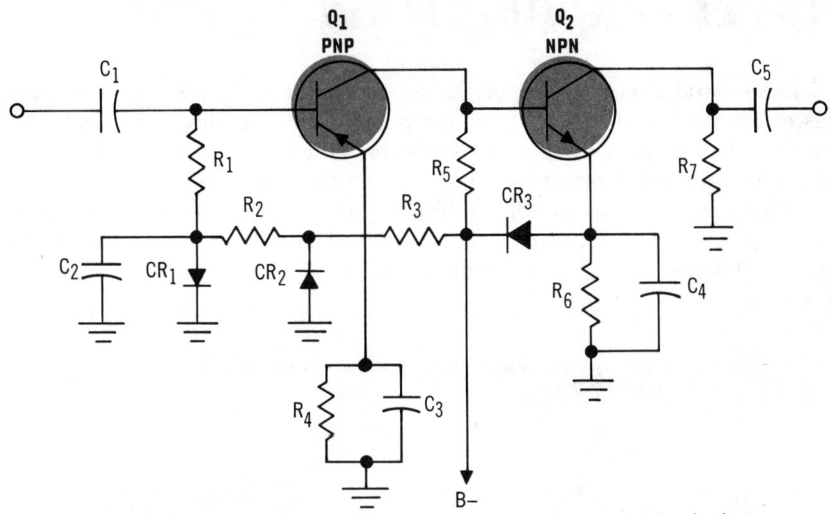

D-c amplifiers have considerable thermal instability and so must use compensating circuits to minimize gain drift

direct coupling

Direct coupling is often used for a good frequency response since it eliminates the inductive or capacitive reactances in the signal path, and so allows very little impedance change with frequency. Direct coupling has disadvantages, though. One is that the base circuits also amplify direct-current signals, and since the bias currents are dc, any changes in d-c bias current will be amplified. Thus, the d-c amplifier is very *temperature sensitive* and needs *stabilizing circuits* to minimize *gain drift*.

Two same-type transistors directly coupled need stepped-up bias voltages to get proper bias currents

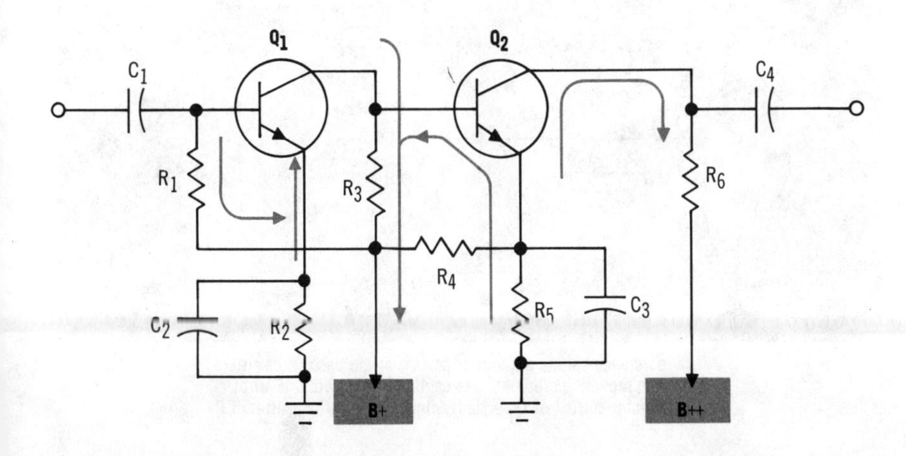

direct coupling (cont.)

The second disadvantage of the *d-c amplifier* is that when both transistor stages are of the n-p-n or p-n-p type, different level bias voltages are needed to provide proper operation, since the input bias of one stage is obtained from the output bias of the previous stage. This can be overcome though, by alternating n-p-n and p-n-p types, since they require opposite polarities. Either a tapped supply can be used to get both polarities, or different arrangements can be used with only one polarity.

When different type transistors are coupled, the power supply voltage requirements become simplified

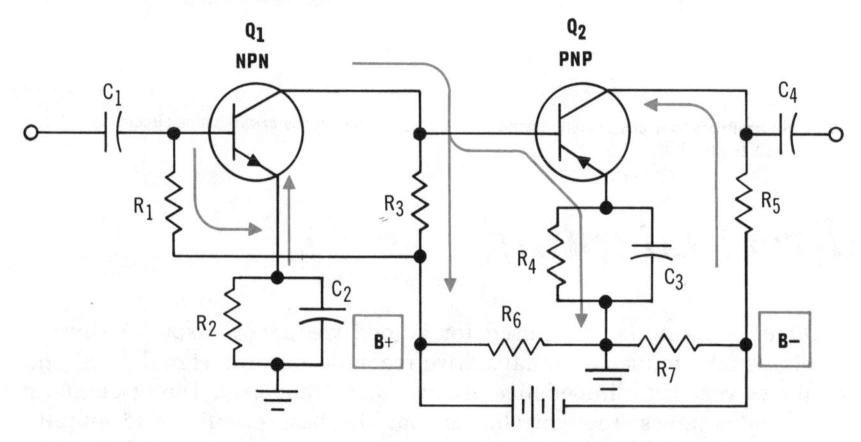

This circuit works the same as the above circuit. The collector of Q_2 is still negative with respect to the base and emitter

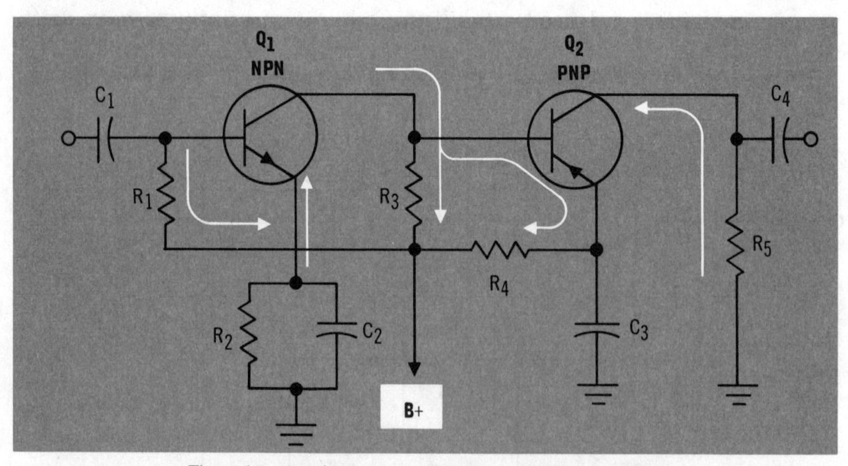

The resistance values are very important in the base and emitter circuit to provide the proper base-emitter forward bias. Part of the collector current of Q_1 actually flows through the emitter of Q_2

photodiodes and phototransistors

Semiconductors work the way they do because current carriers are produced when energy is applied to release electrons from their bonds. This energy is, in effect, producing current carriers. You know that energy can be applied in the form of heat or voltage. Any other form of energy can also be used to produce the same effect, such as *light* or *photon energy*. The *photodiodes* and *phototransistors* operate on this principle.

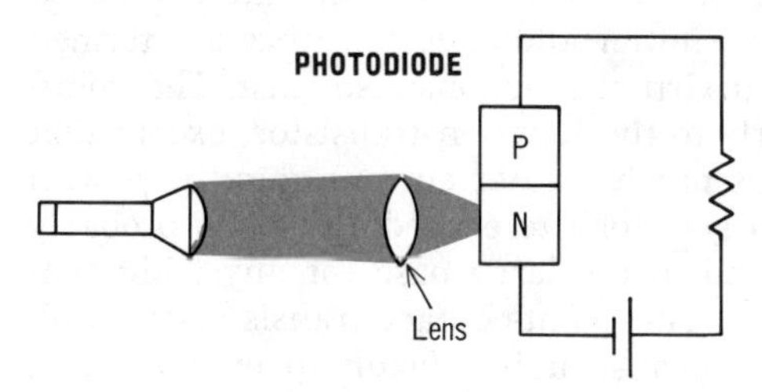

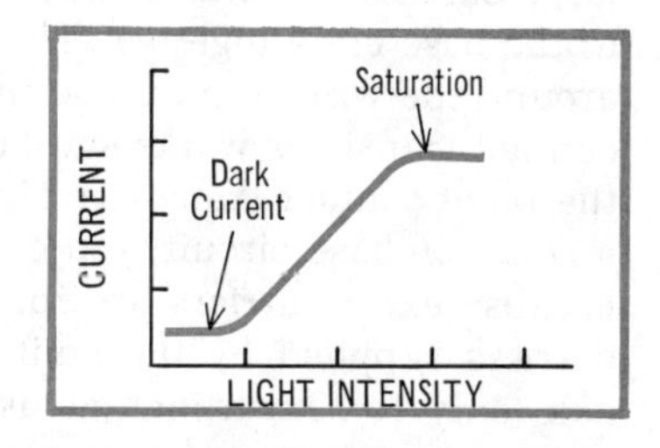

A simple photodiode is connected in a circuit with forward bias with a certain current flow. Then when light strikes one of its elements, the photon energy is given off to the atoms to release more current carriers. This reduces the resistance of the diode, and current flow goes up. When a transistor is made this way, and light energy is applied to the emitter or base, the current changes that result from the light can be amplified in much the same way as in a regular transistor circuit.

With photodiodes and phototransistors, the current that flows when no light is applied is called *dark current;* current will then increase when light is applied, and increase further when the intensity of the light goes up. When the point is reached where a further increase in light intensity does not increase the current further, the device is *saturated.* A lens is used with these devices to focus the light rays on a small area to increase sensitivity.

A special balanced input phototransistor is made with its emitter and collector having the same size and doping, so that it acts like two p-n diodes back to back. There is no amplification. It only *compares* light intensities.

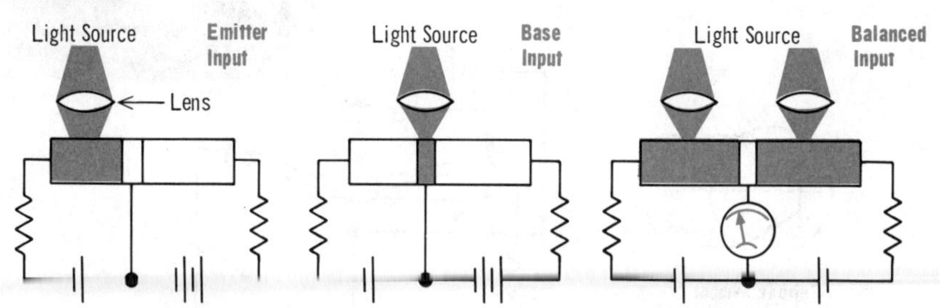

The phototransistor amplifies the effect of the light. The balanced-input phototransistor is specially constructed to compare the intensities of the two light sources

point-contact and tetrode transistors

The transistor theory that you learned was based on the basic grown junction transistor since that is the common type. There are other types, though, which will be described later, that do not function in the exact same manner. However, their *interactions* are essentially similar.

The *point-contact transistor* was the first type made. In this type, a large base material is used, and two "cat's whisker" wires are connected to the base close together. Then, emitter and collector areas are formed around the wire contacts, so that virtual junctions also exist. The point-contact transistor works similarly to the junction transistor, except that the point-contact transistor gives much greater current gain; even with a common-base circuit, you can get alpha to exceed 1. This is probably because extra carriers are formed in the large base for any additional carriers supplied by the emitter. The point-contact transistor is available only for experimental use because it is difficult to manufacture, cannot carry large currents, and is unstable.

The *tetrode transistor* is the same as the junction transistor, except that it provides an extra lead for applying a potential *across the base*. This potential sets up an electrostatic field that attracts and repels the normal current carriers, so that they are concentrated in a smaller area. This has the effect of increasing response time, lowering base resistance, and decreasing junction capacitance, all of which increase frequency response. Since the tetrode potential affects base resistance, it can be varied to control gain, or it can be used as a modulating signal input for heterodyning.

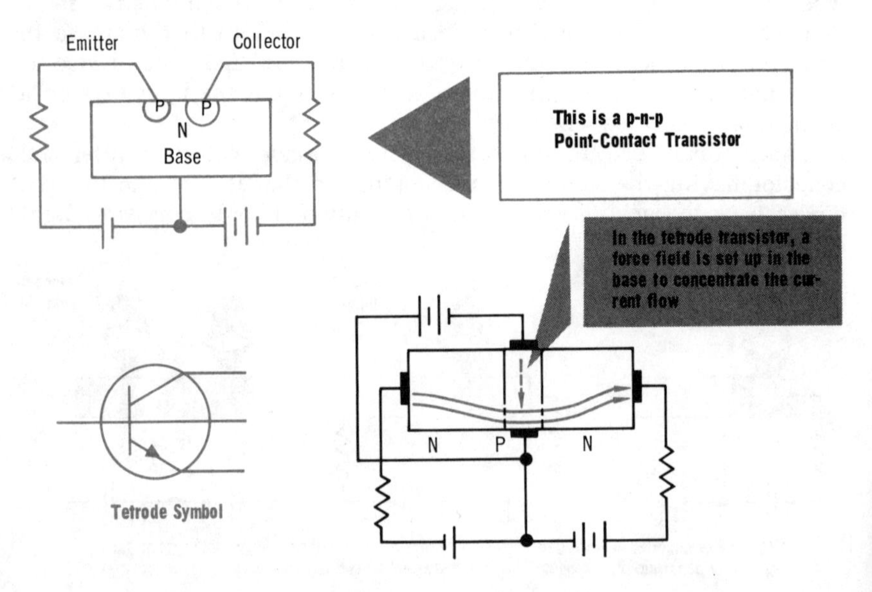

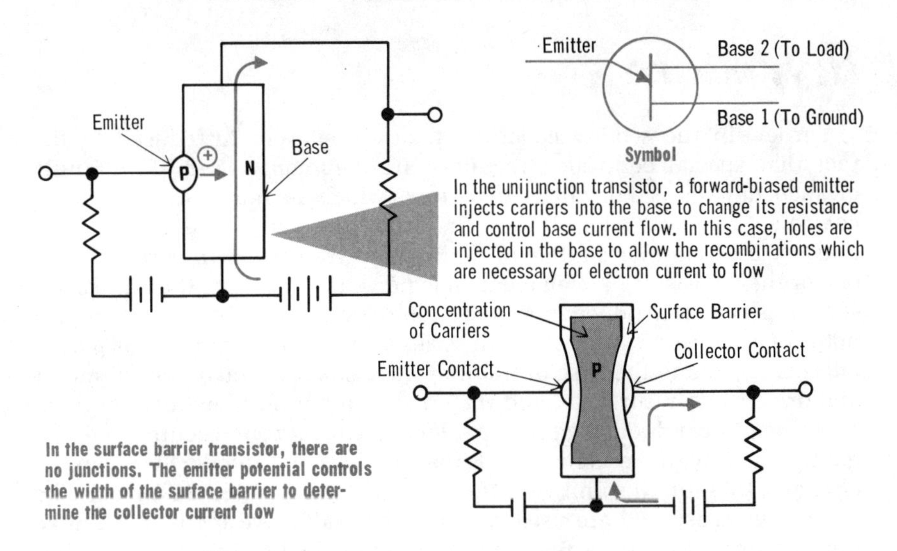

In the unijunction transistor, a forward-biased emitter injects carriers into the base to change its resistance and control base current flow. In this case, holes are injected in the base to allow the recombinations which are necessary for electron current to flow

In the surface barrier transistor, there are no junctions. The emitter potential controls the width of the surface barrier to determine the collector current flow

unijunction and surface barrier transistors

The *unijunction transistor* is more of a *controlled-diode* type of device, although it can amplify. It contains a base slab, and an emitter at its center. There is no collector, and so there is only one junction. A voltage is applied *across* the base, so that at the center, where the emitter is, this voltage is divided in half. The voltage applied to the emitter, then, will determine whether that junction is forward or reverse biased. If a 12-volt battery is applied across the base, and 6 volts is applied to the emitter, then no bias voltage will appear across the junction since the base is also 6 volts at that point. The current flow through the base will be low and only depend on the base battery. But, if the emitter voltage is made +7.5 volts, the junction will be forward biased with 1.5 volts; the emitter will then inject majority carriers into the base to reduce its resistance and increase its conduction. This device is usually used in this way with the emitter input used to gate or trigger conduction through the base. When the opposite ends of the base are connected, the unijunction transistor acts like a simple p-n diode.

The *surface barrier transistor* also uses a large base material, but no real emitter and collector junctions. Instead only contacts are used across a wafer thin section of the base. Its operation depends on a depletion region that develops around the surface of the semiconductor material, as was explained on page 4-30. The current carriers in the base accumulate in a region beneath the surface. A near empty region around the surface acts as a barrier to collector current flow. The varying potential at the emitter has the effect of widening or narrowing the barrier area, through electrostatic attraction and repulsion, to change the resistance to current flow in the collector circuit.

thyristors

Earlier, in the section covering diodes, you were introduced to the fact that special controlled rectifiers are commonly used. The family of such devices is known as *thyristors*, which includes *SCR's* (*silicon controlled rectifiers*), and *triacs*. The SCR's use four elements (p-n-p-n) that act in a manner similar to that of two transistors in series. When one of the transistor base elements has no voltage applied, that transistor section is cut off. And when that section does not conduct, the other will not because the two sections are in series. But when a forward-bias *gating pulse* is applied to the base of the first transistor, it is taken out of cutoff and driven into saturation, and the entire p-n-p-n unit conducts heavily. After heavy conduction starts, *avalanche breakdown* occurs, and the gating pulse can be removed. Either *cathode* or *anode* bias can be changed to shut the unit off. In this way, the p-n-p-n SCR's act like *thyratron* tubes, and are used as controlled half-wave rectifiers. Unlike the thyratron, though, a reverse-bias pulse applied to the gate will shut it down.

The triac is a special bidirectional device that also provides high breakdown currents when a heavy gating pulse is applied to the base. The triac is used in high-power work, and is more efficient than the SCR since it conducts for a full a-c cycle when it is gated.

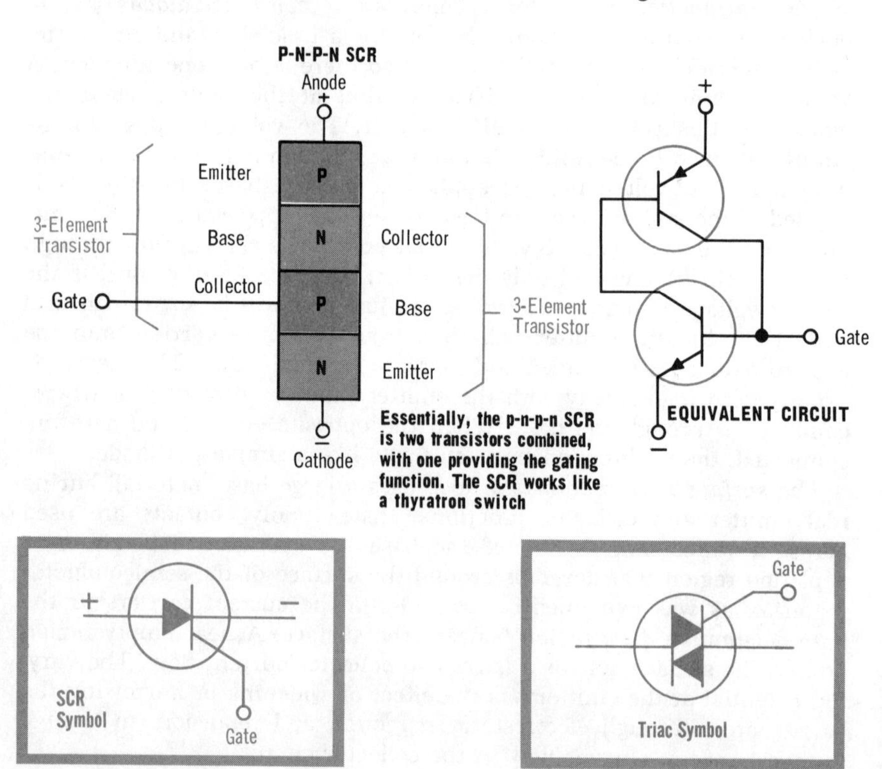

Essentially, the p-n-p-n SCR is two transistors combined, with one providing the gating function. The SCR works like a thyratron switch

temperature considerations

Throughout you have been taught that the transistor and other semiconductor devices are *temperature sensitive*. Within certain temperature ranges, the effect of heat on the current carriers is *reversible*, which means that when the transistor cools again, its characteristics will return to normal. But, above a certain temperature, *irreversible changes* can take place that would make the transistor useless because its characteristics would not return to normal. To avoid these irreversible changes, care must be taken to limit how hot a transistor becomes.

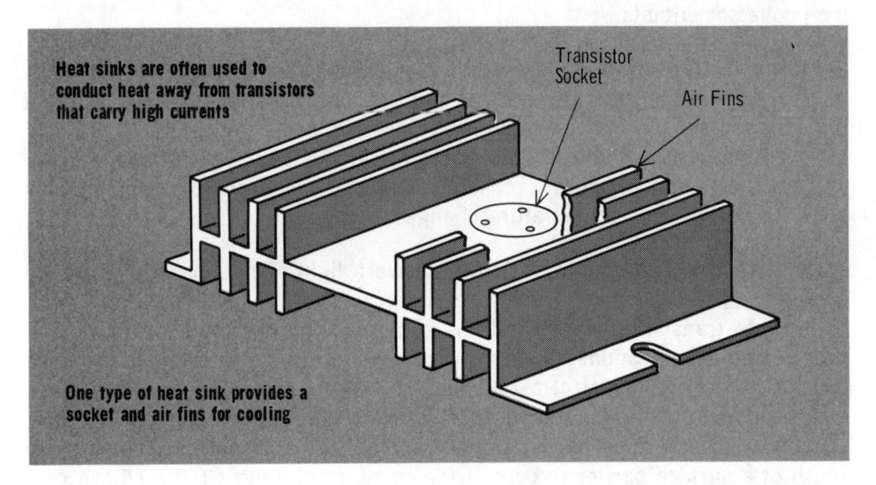

Heat sinks are often used to conduct heat away from transistors that carry high currents

One type of heat sink provides a socket and air fins for cooling

A special cooler-running soldering iron must be used to solder and unsolder transistor leads; and a metal tool, such as a long-nose pliers, must be held on the lead being soldered to conduct the heat away. Transistors should not be mounted near heat-producing parts, and where heat presents a problem, a *heat sink* should be used. Heat sinks are metal parts that a transistor can be mounted on or within, so that the heat will be conducted by the sink. Generally, heat sinks are *lamp black* so that they will radiate their heat to the air.

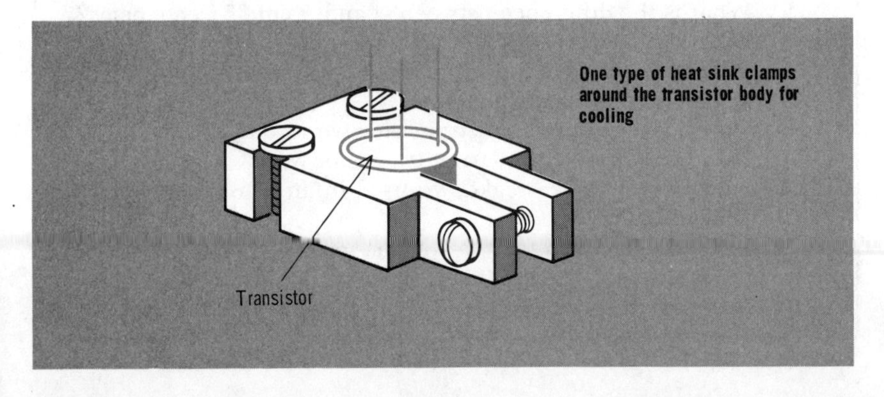

One type of heat sink clamps around the transistor body for cooling

summary

☐ Transistor circuits oscillate by feeding back a portion of the output in phase with the input. ☐ Common-base transistor circuits oscillate readily, since the input and output signals are ordinarily in phase. ☐ For common-emitter transistor circuits to oscillate, they must be fed with a phase-reversed signal from a transistor, since there is a 180-degree phase shift between the input and output circuits. ☐ The classes of operation, A, B, and C, differ from transistors to tubes due to differences in operating characteristics. ☐ Transistor cutoff frequencies are comparable to tubes, and are given in terms of α cutoff for common-base circuits, and β cutoff for common-emitter and common-collector circuits.

☐ Maximum transfer of a signal requires matching the high output impedance of one stage to the low impedance of the next stage. ☐ Transformer coupling provides impedance matching, but has poor frequency response. ☐ RC and LC coupling improves the frequency response, but results in a loss of gain. ☐ Direct coupling provides excellent frequency response, but is extremely sensitive to bias and temperature changes.

☐ Photodiodes and phototransistors convert light variations to resistance changes. ☐ Point-contact transistors work similarly to junction transistors. ☐ Tetrode transistors work similarly to junction transistors but provide improved frequency response due to a concentration of current flow in the base caused by an applied electrostatic field. ☐ The unijunction transistor is a single-junction controlled diode that uses resistance changes in its base section. ☐ The surface barrier transistor amplifies a signal by changing the width of a surface barrier to control the collector current. ☐ The silicon controlled rectifier is a four-element p-n-p-n device that acts like a thyratron tube. ☐ The triac is a bidirectional gate.

review questions

1. Why do common-base circuits readily oscillate?
2. What is the difference between α and β cutoff frequencies?
3. Why is impedance matching important?
4. Why is direct coupling of transistor stages used?
5. What is the *tetrode transistor*?
6. How does a unijunction transistor work?
7. How does the surface barrier transistor work?
8. How is avalanche breakdown used in an SCR? How does a triac differ?
9. Draw the symbols for the transistors and controlled diodes of Questions 5 through 8.
10. What is a heat sink?

the field-effect transistor (FET)

The *field-effect transistor*, more commonly called the *FET*, functions differently from the ordinary junction transistor, yet operates on some of the same principles. The regular junction transistor is called a *bipolar* transistor because it uses two current carriers, both holes *and* electrons. The FET uses only one current carrier, either holes *or* electrons, depending on whether it is a p- or n-type; hence the FET is *unipolar*.

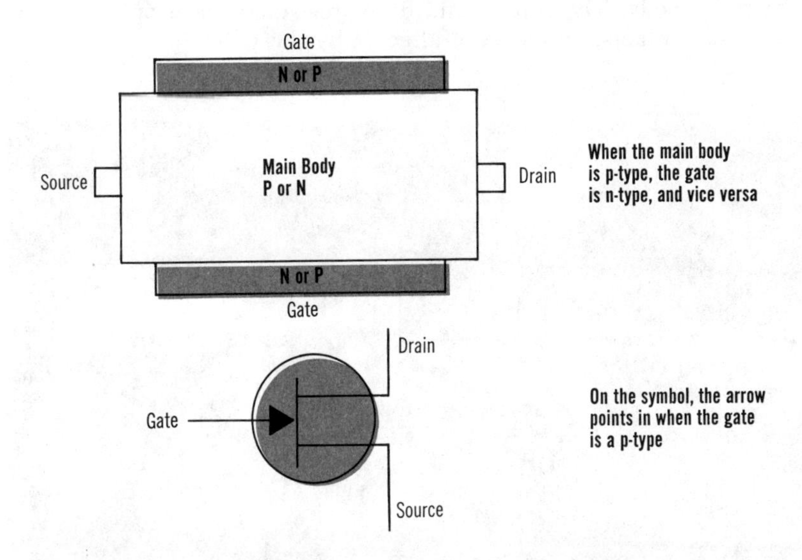

The FET has a high input resistance, similar to the electron tube, and unlike the low input resistance of the junction transistor. The FET is a signal-*voltage*-sensitive unit, whereas the junction transistor is *current* sensitive. As a result, the FET can work better with low-voltage, low-power inputs. The overall advantages of the FET are: (1) high input impedance, (2) good input/output isolation, (3) low-signal voltage sensitivity, and (4) high gain. It has transconductance properties similar to the electron tube, and needs only low power supply voltages similar to the junction transistor. There are two basic types of FET's: the JFET and the MOSFET. The JFET (junction FET) will be used for the basic theory; then both will be described in more detail later.

The FET uses a *source*, *gate*, and *drain*, as opposed to the emitter, base, and collector of the junction transistor. The source supplies the current, the gate controls the current, and the drain collects it. The source and drain are merely contacts at opposite ends of the main body, which can be either a p- or n-type. The gate forms a junction around the main body, and is n-type when the main body is p-type and vice versa.

input/output circuit interaction

The FET transistor amplifies in a manner very similar to that of the electron tube, and somewhat like the junction transistor. It is set up with two circuits: an *input circuit*, and an *output circuit*. The two circuits are arranged so that they interact. The *voltage* that is applied to the input circuit controls, to a large extent, the *current* that flows in the output circuit. Therefore, when a signal is applied to the input circuit, it produces a corresponding variation in the output current. This is accomplished by the signal voltage on the gate controlling the resistance of the main body. The source and drain resistances are purely ohmic due to lead contacts, and are not affected by the gate.

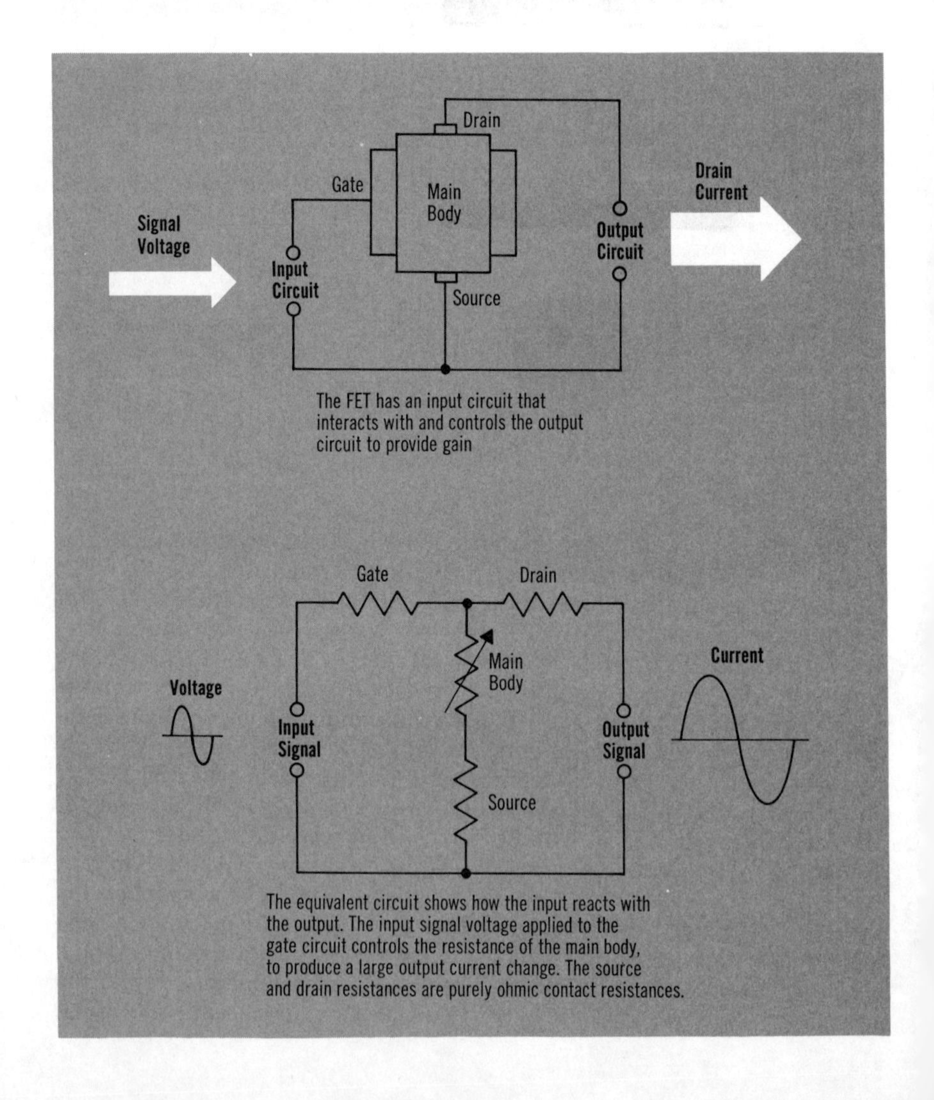

The FET has an input circuit that interacts with and controls the output circuit to provide gain

The equivalent circuit shows how the input reacts with the output. The input signal voltage applied to the gate circuit controls the resistance of the main body, to produce a large output current change. The source and drain resistances are purely ohmic contact resistances.

the FET main body

The FET main body is merely a semiconductor block doped for either electron or hole majority carriers, depending on whether an n or p unit is used. The block is made relatively large, compared to the gate, and is sufficiently doped to provide the number of ***majority*** carriers needed for proper current flow. As with any other semiconductor device, the doping and the size are designed to provide for the specified current. Also, because of the way in which a FET works, which will be taught later, additional doping is used to give the block a *large number* of *minority* carriers as well. This is similar to what is done with the collector of a junction transistor, as explained on page 4-55.

The main body, by itself, is merely a semiconductor *resistor*, whose resistance is determined by the number of carriers available. The *source* and *drain* leads are attached to opposite ends of the main body with *ohmic* connections; that is, no p-n junctions are formed, so that the leads only conduct current and do not play an active role. When a voltage is applied across the source and drain leads, current flows through the main body as explained on page 4-18. If the main body is *symmetrical*, either end of it can be the source or drain. If it is *unsymmetrical*, the source and drain cannot be interchanged. However, in either case, the symmetry of the gate plays an important role.

The path that the current takes through the main body is called the *channel*. An important effect in the FET is that current flow through the channel causes discrete voltage drops across segments of the channel, as though the channel were comprised of a series string of discrete resistances. This produces a *voltage gradient* across the channel. This effect occurs in all resistances that conduct current. But in the FET, it is important because of its *interaction* with the *gate*. This is similar to what was explained for the unijunction transistor on page 4-109.

The main body is relatively large and is doped to produce a large number of majority and minority carriers

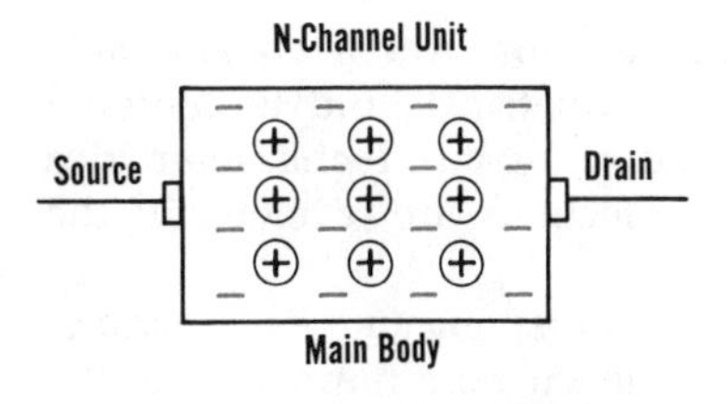

If the main body is symmetrical, the source and drain connections can be interchanged

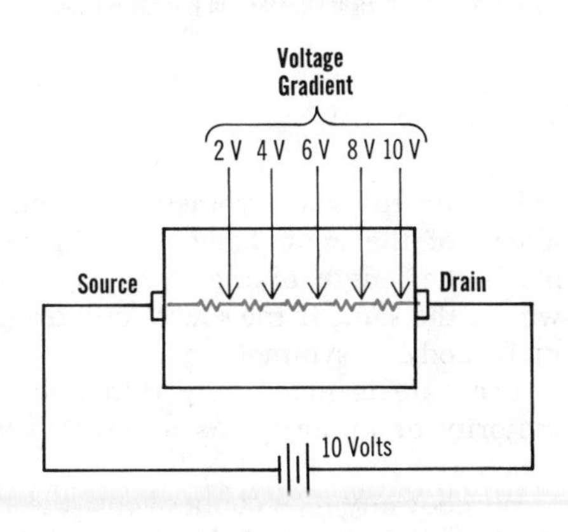

Current flowing through any semiconductor or conductor produces a voltage gradient, a rising voltage drop across the unit

the FET gate

The FET *gate* is used to control the current flow through the channel. The gate forms a p-n junction with the main body, and is doped to have majority carriers opposite from it. The gate can be made to encircle the main body completely, or it can merely be on opposite sides of it.

A gate that forms a *continuous p-n* junction around the main body requires only one connection. The *two-gate* unit must be externally connected, or both gates can be excited as separate current controls. Generally, the continuous-gate has greater current control because of the larger p-n junction area.

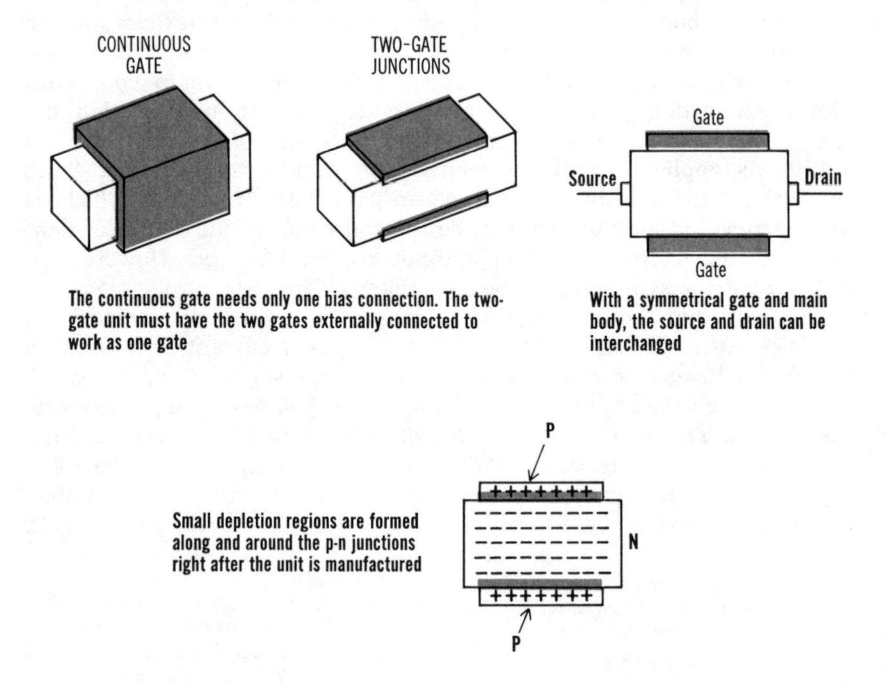

The continuous gate needs only one bias connection. The two-gate unit must have the two gates externally connected to work as one gate

With a symmetrical gate and main body, the source and drain can be interchanged

Small depletion regions are formed along and around the p-n junctions right after the unit is manufactured

Usually the gate geometry is such that it runs over a considerable length of the main body. If the gate is symmetrical, and is centered along the length of the main body, its current-control characteristics will be the same if the source and drain connections are reversed (if the main body is symmetrical).

The gate is made very thin and so contains *few current carriers*, majority or minority. As a result, very little current flows across the junction.

Without any bias connections, the majority carriers on either side of the junction attract each other and combine to form a depletion region, the same as explained on page 4-30. The depletion region, and how it is controlled, determine the operation of the FET.

FET channel voltage

When the gate bias is not varied, the voltage applied to the source and drain contacts determines the *channel current.* The current will rise as the source/drain voltage is increased, until at a certain voltage the channel current no longer increases. This is the *pinch-off voltage*, and is similar to saturation current in a junction transistor.

Pinch-off occurs because a *voltage gradient*, developed along the main body, affects the *depletion region.* On the previous page, the depletion region was shown without applied voltages, and without a voltage gradient.

With zero bias at the gate, and a voltage between the source and drain, current will flow through the channel to set up a voltage gradient. This voltage gradient exists at the gate, all along the length of the p-n junction. With the n-type FET shown, the rising positive gradient along the main body results in an apparent rising negative *reverse-bias gradient* along the gate. The negative bias attracts the *minority* carriers (holes) in the main body close to the gate, causing accumulations of the carriers in proportion to the voltage gradient. These accumulations of holes attract the negative majority carriers to fill them, *depleting* those *regions* of carriers and effectively narrowing the *conduction channel.* Since the shape of the channel is no longer symmetrical, neither is its resistance, and hence, neither is the voltage gradient. As the source/drain voltage is increased, there are enough majority carriers to allow increased channel current. The gradient voltages closer to the source go higher, increasing the area of the depletion region until *pinch-off* is reached, where current no longer rises.

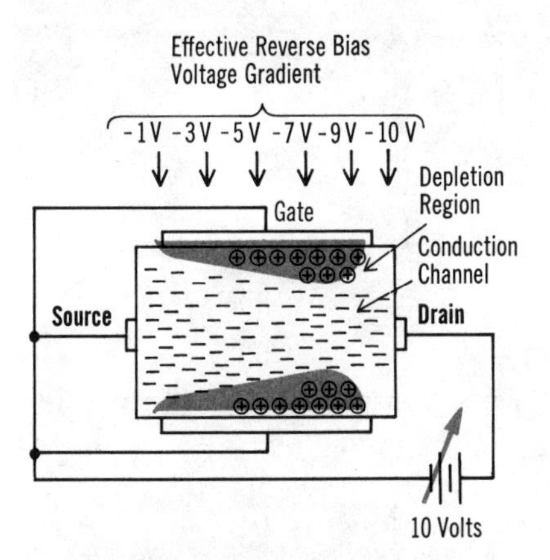

The voltage gradient along the main body results in a reverse bias voltage gradient along the gate which enlarges the depletion region

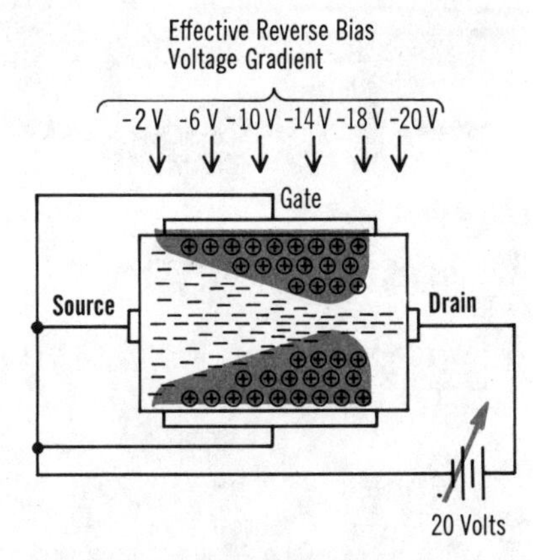

As the source/drain voltage is raised, channel current rises, but the voltage gradient changes to increase the depletion region and narrow the conduction channel. This will continue until pinch-off is reached, where the channel current will no longer rise

FET bias

Since the level of source/drain voltage results in an effective reverse bias gradient along the gate, the overall operation of the FET is determined by the interaction of the source/drain voltage and an applied *gate bias*. Actually, the voltages are chosen so that changes in gate bias voltage will bring about a voltage range in which the depletion region will be reduced to some current saturation level, and increased to a current cutoff level. However, unlike the effects of varying the source/drain voltage, the current that flows in the conduction channel falls as the reverse bias is increased, and vice versa. An increase in reverse bias draws more minority carriers out of the conduction channel to attract and neutralize majority carriers, thus enlarging the depletion region. The reduction in the available number of majority carriers causes the reduction of channel current. The depletion region encompasses the area of reduced carriers. The conduction channel signifies the area where majority carriers, though reduced in number, are still available to conduct current. When the reverse bias is increased to where the depletion region closes the conduction channel, there are no majority carriers available where the channel is closed to conduct current, and cutoff occurs.

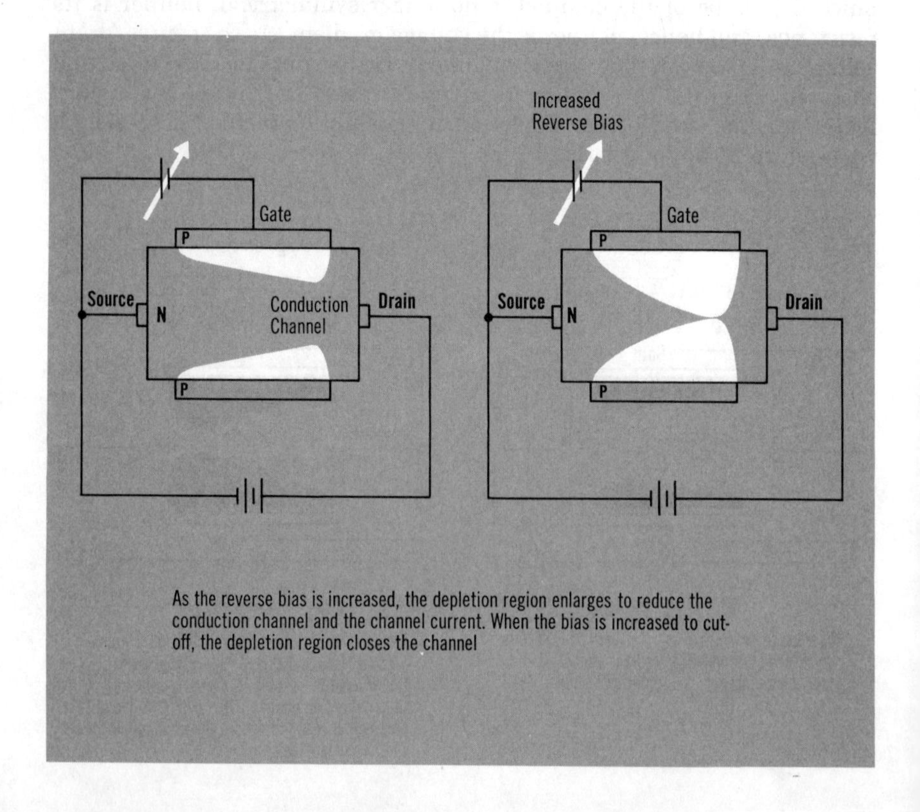

As the reverse bias is increased, the depletion region enlarges to reduce the conduction channel and the channel current. When the bias is increased to cutoff, the depletion region closes the channel

FET signal operation

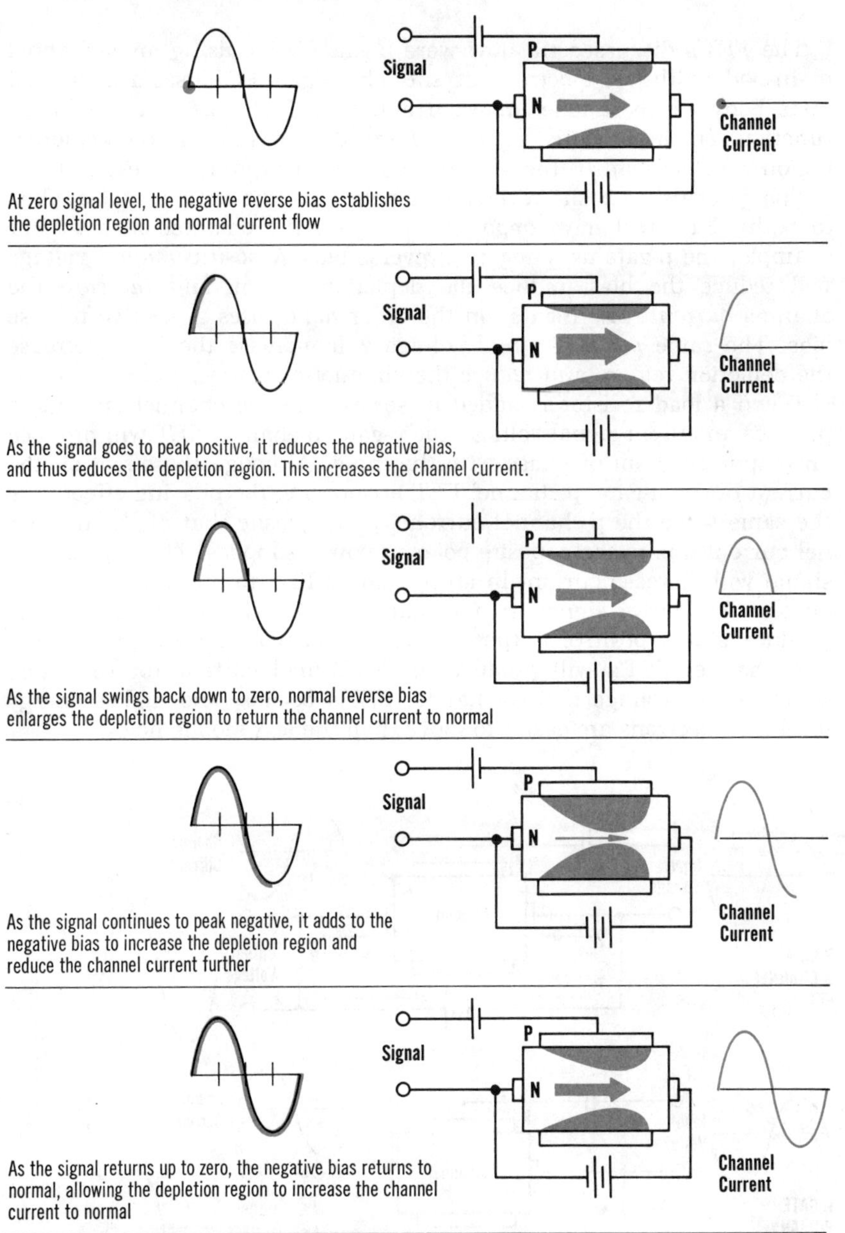

At zero signal level, the negative reverse bias establishes the depletion region and normal current flow

As the signal goes to peak positive, it reduces the negative bias, and thus reduces the depletion region. This increases the channel current.

As the signal swings back down to zero, normal reverse bias enlarges the depletion region to return the channel current to normal

As the signal continues to peak negative, it adds to the negative bias to increase the depletion region and reduce the channel current further

As the signal returns up to zero, the negative bias returns to normal, allowing the depletion region to increase the channel current to normal

The signal voltage is applied in series with the reverse bias. As the signal voltage swings, it adds to or subtracts from the bias, thus varying the overall bias. This causes the depletion region to vary, changing the channel current so that it follows the signal voltage.

n- vs. p-gate FET's

The FET's discussed till now were p-gate FET's, using an n-channel main body with *free electron* current. The n-gate FET uses a p-channel main body which supplies *hole* current. Both FET types are similar in function. The input signal modifies the reverse bias to vary the depletion region, and so control the channel current. However, since the p- or n-type gates use opposite reverse bias polarities, the same signal applied to both FET's will have *opposite effects* on the channel *current*. For example, the p-gate uses negative reverse bias. A *positive signal* voltage will reduce the bias, reduce the depletion region, and *increase* the channel *current*. The n-gate, on the other hand, uses a positive reverse bias. The same *positive signal* voltage will increase the bias, increase the depletion region, and *reduce* the channel *current*.

When a load resistor is added in series with the channel current to produce an output signal voltage, the p-gate, n-channel FET will produce an output 180° out of phase with the input. Although the effect on the current of the n-gate, p-channel FET is opposite, the phasing effects are the same since the p-channel current runs opposite that of the n-channel current due to the opposite polarity power supplies. Thus, a positive signal will increase current in an n-channel FET to produce a negative output. A positive signal in a p-channel FET will reduce current to produce a less positive output, which is a negative swinging output.

N-channel FET's will produce more channel current for the same source/drain voltage than p-channel FET's because the higher energy level free electrons are easier to move than the less mobile holes.

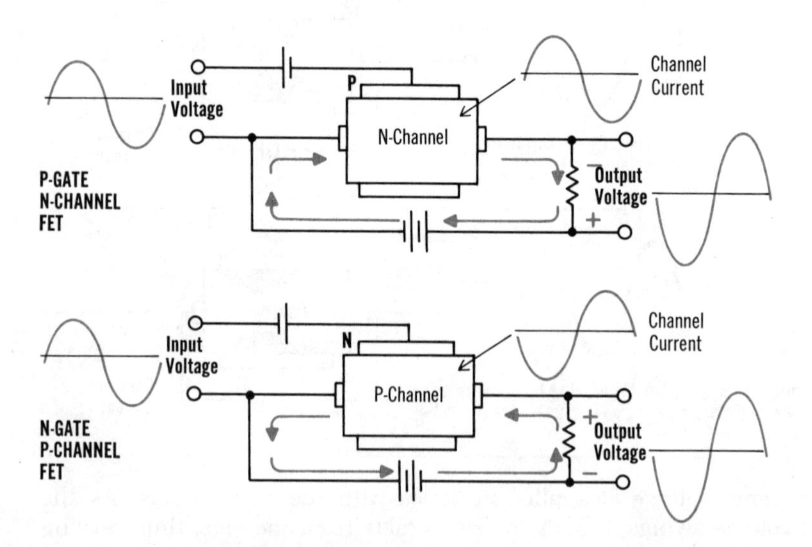

Increased current in an n-channel FET produces a negative output. Reduced current in a p-channel FET produces a less positive output, which is the same thing

FET transconductance and gain

The FET works in a manner similar to that of the electron tube in that it uses an *input voltage* to control an *output current*. As with the electron tube, *transconductance* is the characteristic that signifies the amount of control that the input voltage has over the current. Transconductance is expressed in micromhos, and is found with:

$$g_{fs} = \frac{\Delta i_D}{\Delta e_G} \times 1000$$

where:

g_{fs} = transconductance in micromhos
Δe_G = a change in gate/source voltage in volts
Δi_D = the resulting change in drain current in ma

The transconductance of a FET will vary with the type of FET used, as well as with the current and voltage parameters of the circuit design. Typical transconductance values range from under 50 micromhos to over 50,000 micromhos.

Since transconductance is the expression of how much control the gate voltage has over drain current, it also determines the amount of *voltage gain* available from a FET when the change in drain current produces a change in output voltage across a load resistor, R_L. Voltage gain is found with:

$$A = \frac{\Delta e_o}{\Delta e_i}$$

where:

A = voltage gain
Δe_o = change in output voltage
Δe_i = change in input voltage (gate/source voltage)

For example, an input voltage change of 0.1 volt which causes an output voltage change of 3 volts would result in a gain of 30. Since the voltage gain is dependent on the values of transconductance and the output load resistor, it can also be calculated from:

$$A = g_{fs} R_L$$

A load resistor of 5000 ohms and a transconductance of 6000 micromhos will produce a gain of 30 ($0.006 \times 5000 = 30$).

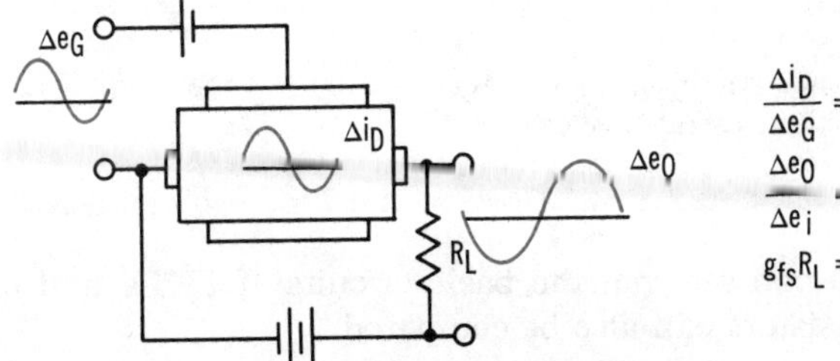

$\frac{\Delta i_D}{\Delta e_G} = g_{fs}$ (Transconductance)

$\frac{\Delta e_o}{\Delta e_i}$ = A (Voltage gain)

$g_{fs}R_L$ = A (Voltage gain)

comparison of FET's, junction transistors, and tubes

Just as the junction transistor was compared to the electron tube earlier, the FET can be compared to both, and more appropriately.

The drain in the FET collects the current, as do the plate in the electron tube and the collector in the junction transistor. In the FET, the current is supplied by the source, which is similar to the current being supplied by the cathode in the tube and the emitter in the junction transistor. The gate in the FET controls the current, as do the control grid in the tube and the base in the junction transistor.

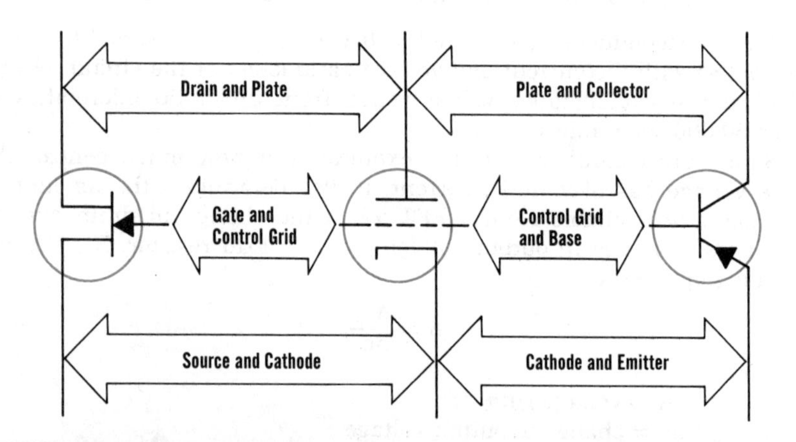

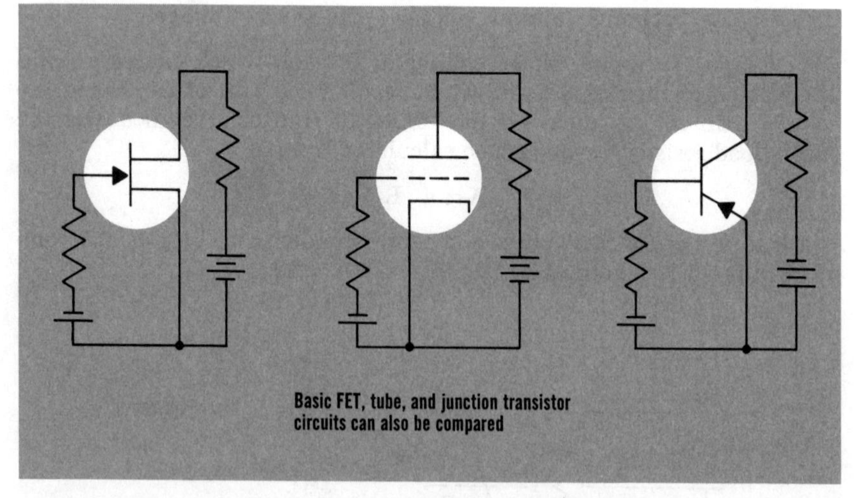
Basic FET, tube, and junction transistor circuits can also be compared

From the above, you can see that the basic circuits of FET's, and of tube and junction transistors can also be compared.

JFET's

Unlike other junction transistors, which rely on forward biased input circuits, FET's have a high input impedance similar to electron tube circuits, and so there is less difficulty in *impedance matching cascaded* FET stages. In addition, since the main signal path is not *through* a biased junction, there is less difficulty with junction capacitances interfering with high frequency operation. In general, FET amplifiers are better for high frequency work, and are easier to design and use in conjunction with electron-tube circuits.

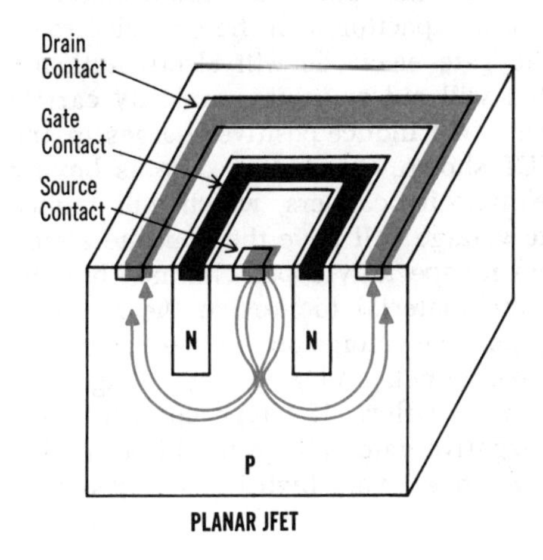

The current from the source contact is channeled through the gate, and is controlled by depleted carriers. The current then divides around the gate and is collected by the drain contact

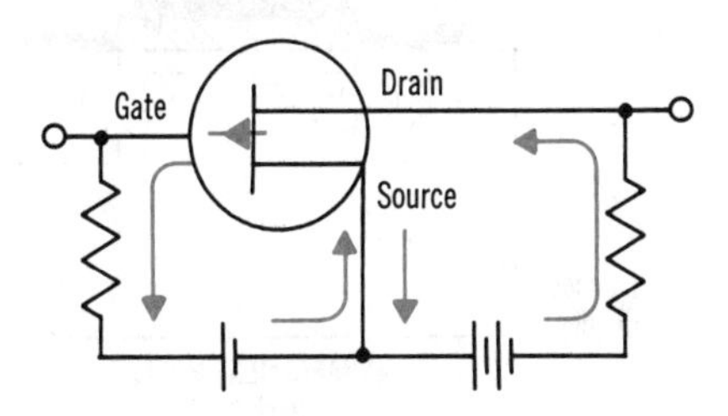

The arrow in a FET symbol shows the doping of the gate and indicates the doping of the channel. The arrow that points into the channel, as shown on the previous page, is a p-type gate, and the symbol is for an n-channel FET. The arrow that points away from the channel, as shown on this page, represents an n-type gate, and the symbol is for a p-channel FET

The planar JFET (junction FET) is manufactured in a slightly different process from the basic grown JFET, but functions in a similar manner. Because of the planar geometry, the current from the source contact is channeled between the gate, and controlled by depleted carriers in much the same way as the grown junction FET. The current that manages to get through the channel divides around the gate and is collected by the drain contact.

MOSFET's

MOSFET, which is the abbreviated name for *metal-oxide semiconductor* field-effect transistor, is the other FET family. These FET's differ from the JFET's in that the gate does *not* form a p-n junction with the channel. Instead, the MOSFET uses a *metal* electrode gate and an *oxide* dielectric to electrostatically control the carrier flow in the *semiconductor*. This electrostatic gate control permits an even higher *input impedance*. The MOSFET is also called an *insulated-gate FET*.

There are two basic types of MOSFET's: the *depletion* MOSFET, and the *enhancement* MOSFET. The depletion type uses a narrow *doped* channel to conduct the current between the source and drain contacts. This channel forms one electrode of a capacitor with the gate electrode, and so any voltage applied to the gate electrode will electrostatically induce charges in the channel that will aid or inhibit majority carrier flow. A negative voltage on the gate will induce positive charges in the channel. In the n-channel MOSFET shown, this is the same as having the minority carriers *deplete* the majority carriers, which will reduce current flow. A positive-going gate voltage will have the opposite effect.

The enhancement MOSFET uses no specially doped channel, but the gate voltage *induces* charges in the material to provide the majority carriers needed for current flow. A positive voltage on the gate, as shown in this example, will induce an n-channel, and a higher voltage will widen the channel and enhance current flow. If p-type contacts were used for the source and drain, a negative gate voltage would be needed to induce a p-channel. This MOSFET has a very high input impedance.

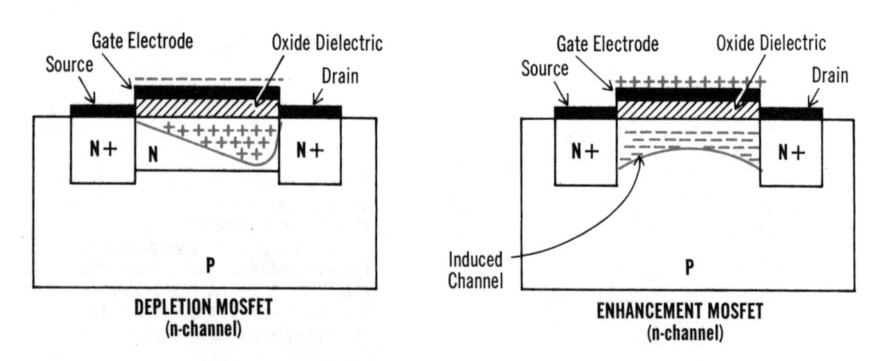

In the depletion MOSFET, the gate voltage induces minority carriers in the channel to deplete their majority carriers and control current flow. In the enhancement MOSFET, the gate voltage induces majority carriers between the source and drain to provide a current channel. The N+ segments provide low resistance contacts for the channel

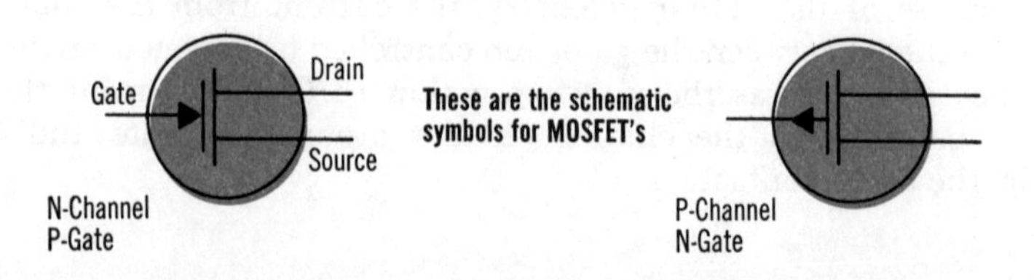

summary

☐ Junction transistors are bipolar since they use two current carriers. FET's use one current carrier, and are called unipolar. ☐ The FET is voltage sensitive, and has a high input resistance similar to the electron tube. ☐ The FET uses a source, drain, and gate. The source and drain are on the main body and thus provide a current channel between them. The source provides the current carriers, and the drain collects the current.

☐ The gate controls the current flow between the source and drain by creating a depletion region in the current channel that is in proportion to the reverse bias of the gate. The more the bias, the greater the depletion region, and the lower the channel current to the drain. ☐ Pinch-off voltage is the voltage at which an increase in source/drain voltage will no longer increase drain current. ☐ Cutoff bias is reached when the depletion region closes the conduction channel. ☐ The respective currents in p- and n-channel FET's are affected oppositely by the same signal voltage, but due to opposite current flow, output phases are the same in both types. ☐ Transconductance is the characteristic that tells how much control the gate has over drain current, and directly affects the gain of the FET. ☐ There are two main classes of FET's: JFET's and MOSFET's.

review questions

1. Why is a FET called unipolar?
2. What are the functions of the source, drain, and gate?
3. What is a depletion region?
4. How does the voltage gradient affect it?
5. How does the input signal voltage affect the depletion region? The drain current?
6. What is pinch-off voltage? Cutoff voltage?
7. What is the equation for transconductance?
8. Does a FET have a high or low input impedance?
9. Does an n-channel FET have an n- or p-gate?
10. Name the two kinds of MOSFET's.

common-source circuit

As shown, the *common-source* circuit is the circuit you have been studying all along. It is so called because the source is common to both the gate and drain circuits (the input and output circuits). The common source circuit is similar to the *common-emitter* circuit of the junction transistor and the *common-cathode* circuit of the electron tube.

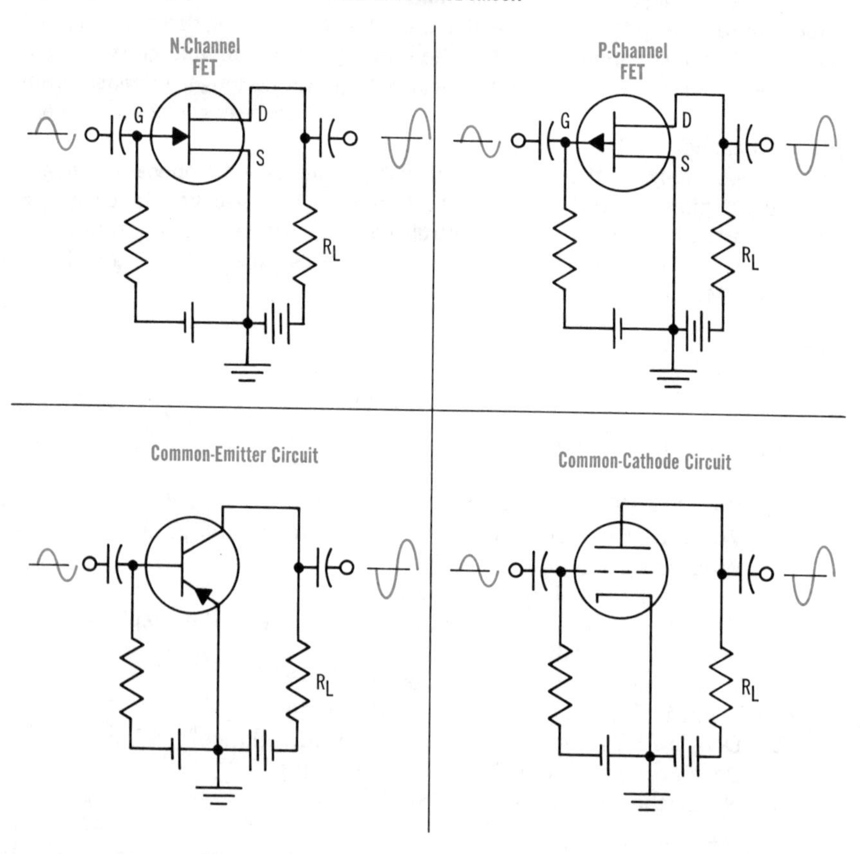

The input signal is applied to the gate, and the output signal appears across R_L at the drain. As the signal goes positive, it opposes the gate bias in the n-channel FET, and increases drain current. The increased drop across R_L causes the drain voltage to go negative, producing a 180° signal phase reversal. With the p-channel FET, the positive signal aids the reverse bias to reduce drain current. Because of the source/drain battery polarity, the lower drop across R_L makes the drain voltage more negative, again producing a 180° phase reversal.

common-gate circuit

The *common-gate* circuit has the gate common to both the source and drain circuits. It is similar to both the *common-base* junction transistor circuit, and the common- or *grounded-grid* tube circuit.

In the common-gate circuit, the input signal is applied to the source. In the n-channel FET, a positive-going signal aids the reverse bias to decrease drain current. The voltage drop across R_L goes down, thus increasing the drain voltage to produce a positive-going output signal. The output phase, then, is the same as the input phase. Actually, this circuit functions the same as the common-source circuit, since a positive signal at the source is the same as a negative signal at the gate, which would also produce a positive-going output signal.

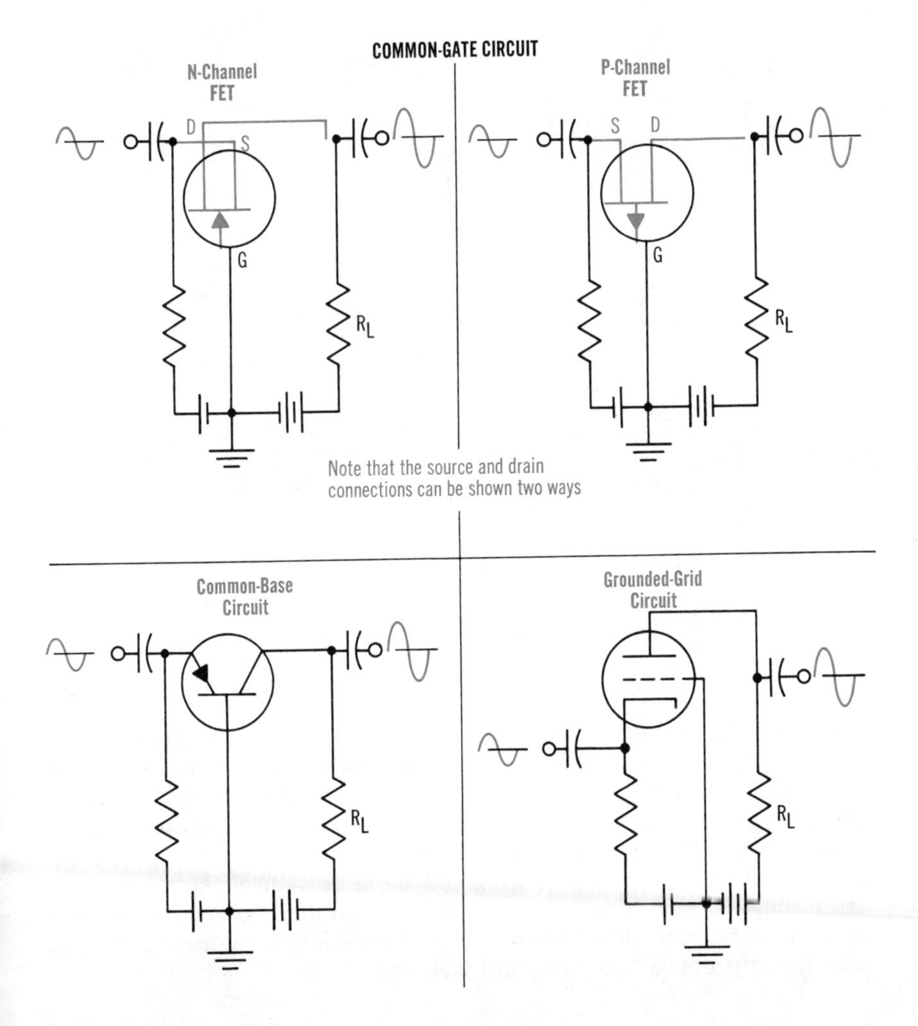

common-drain circuit (source follower)

In the *common-drain* circuit, the drain is common to both the gate and source circuits. This circuit is also called a *source follower* because the output signal, which is taken off the source, follows the input signal in amplitude and phase. The source follower is similar to the *cathode-follower* tube circuit, which is a common- or grounded-plate circuit, and the *emitter-follower* junction circuit, which is a *common-collector* circuit.

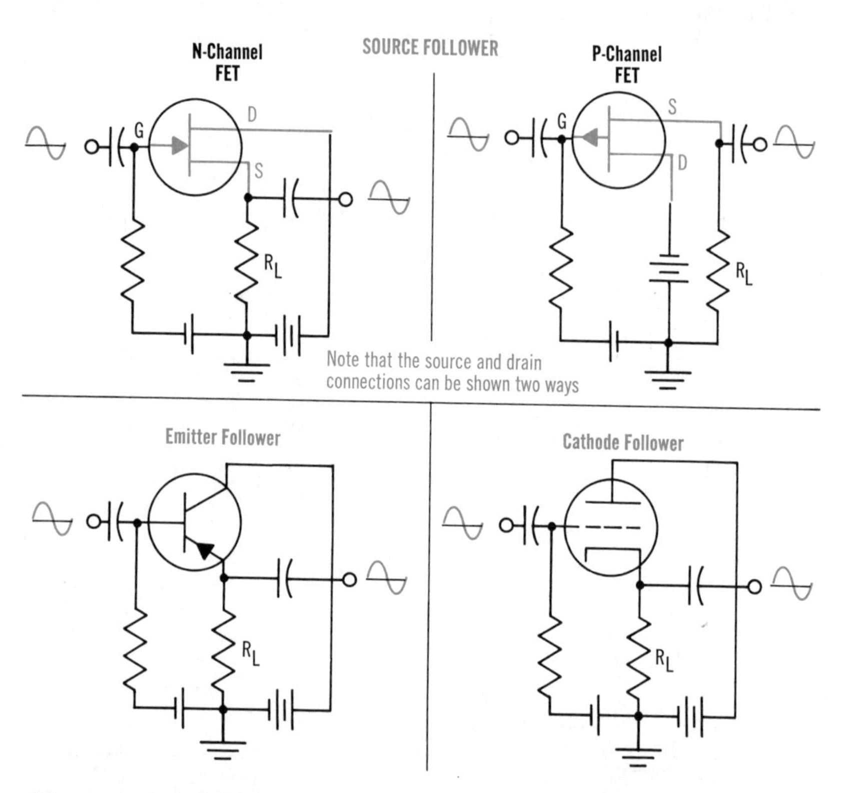

The input signal is applied to the gate, and the output signal appears across R_L in the source circuit. In the n-channel FET, a positive-going input signal opposes the reverse bias, increasing channel current through load resistor R_L. This produces a positive-going output signal. Because the output signal raises and lowers the source voltage in the same direction as the signal voltage on the gate, it degenerates the signal voltage so that there is no gain. The output closely follows the input. Like its cathode-follower and emitter-follower counterparts, it is good for impedance matching and isolation.

operating curves

As with the junction transistor, the FET is manufactured to have certain channel current *characteristics* that are determined by the applied gate voltage, as well as the source/drain voltage. For a single fixed value of reverse gate bias, a single curve will result. This shows how the drain or channel current varies as the source/drain voltage is increased.

As the source/drain voltage is first applied, the channel current rises almost in proportion to the source/drain voltage. Because of this, the first part of the curve is called the *ohmic region*. It is best not to operate the FET in this region because the channel current should be independent of source/drain voltage fluctuations. The flat region of the curve shows that as the source/drain voltage is increased further, the channel current remains almost constant, or *saturated*. This is called the *pinch-off* region. This is the best region for FET operation since the channel current is independent of source/drain voltage fluctuations. Only gate voltage changes will alter the channel current. The end region (shown with dashed lines) is where too high a source/drain voltage would produce *avalanche breakdown*, which could destroy the FET if it were not designed to handle such currents. This is true of all semiconductor devices.

When different values of gate voltage are used, a *family* of channel current curves is produced that shows how the source/drain voltage affects current. The family of curves is used to establish operating conditions for the FET similar to those of the junction transistor (pages 4-75 through 4-79).

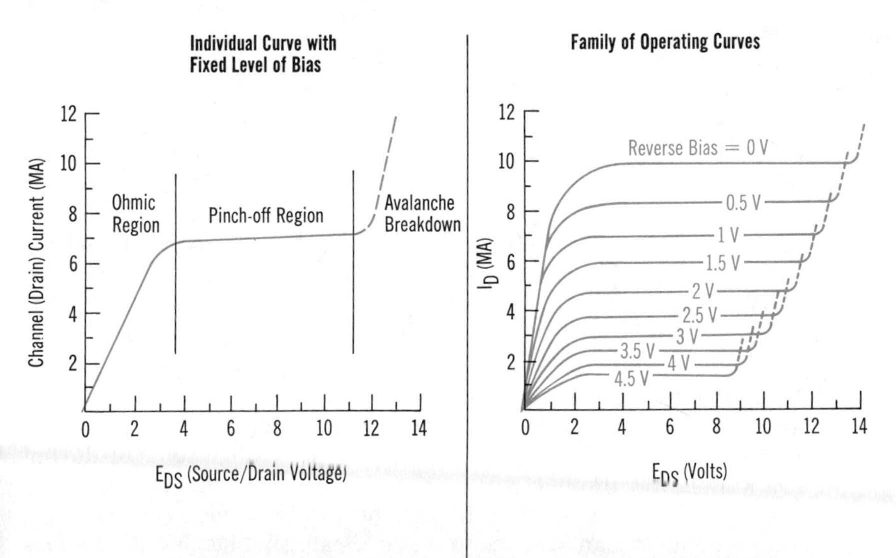

Operation in the pinch-off region will make the drain, or channel, current independent of normal source/drain voltage fluctuations

FET bias methods

In all the previous FET lessons, batteries were used to establish the gate bias voltage. As with the junction transistor, other circuit methods can be used. But, unlike the junction transistor, the FET reverse bias currents are extremely small, and provide no options for controlling bias. The junction transistor is forward biased and can rely on the use of current limiting resistors in a variety of circuit configurations to establish bias. The FET, on the other hand, is similar to the electron tube, and so is biased similarly.

The use of a bias battery is called *fixed* bias, and the value of the battery or a comparable supply voltage sets the bias level. *Source bias*, which is similar to cathode bias with an electron tube, requires no battery. The channel current flowing through the source resistor R_s will produce a voltage drop to bias the gate. For a particular current level, the right resistance value will produce the required IR drop.

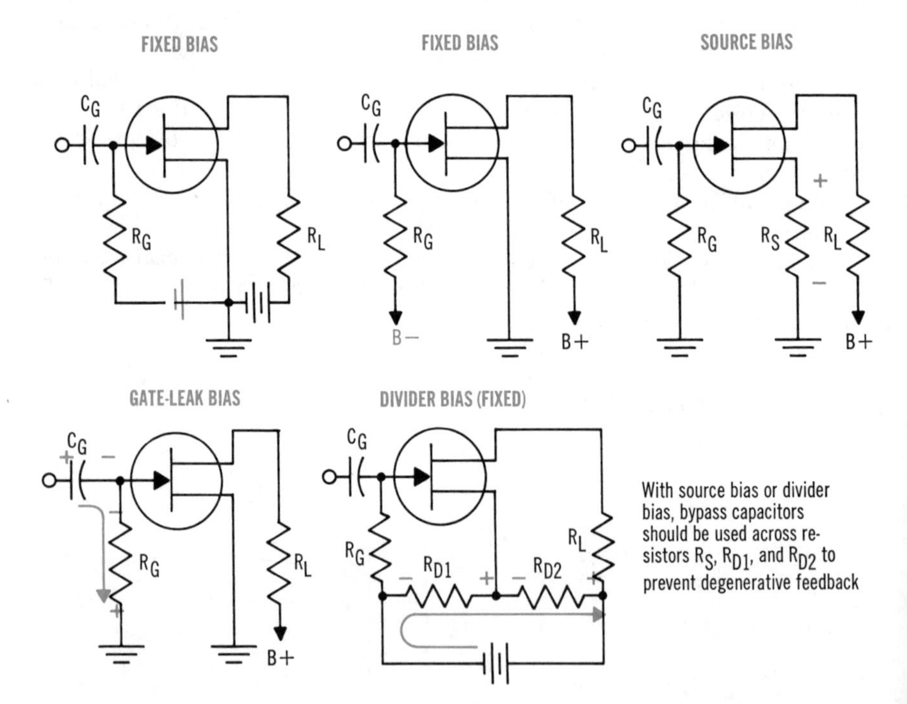

Gate-leak bias, similar to grid-leak bias with an electron tube, will produce a bias when the tip of the input signal forward biases the gate to draw current and charge capacitor C_G, which will discharge during the rest of the cycle through gate resistor R_G. The FET must be designed to handle forward bias currents for this bias to be used.

Voltage dividers can also be used to establish bias levels. This is another form of fixed bias.

summary

☐ The common-source circuit has the input signal applied to the gate, and the output signal taken from the drain. It is similar to the common-emitter and common-cathode circuits. It produces a 180° reversal. ☐ The common-gate circuit has the input signal applied to the source, and the output signal taken from the drain. It is similar to the common-base and grounded-grid circuits. The output phase is the same as the input phase. ☐ The common-drain circuit, also called the source follower, has its input signal applied to the gate, and the output signal taken from the source. It is similar to the emitter-follower and cathode-follower circuits. It produces no gain or phase reversal.

☐ The FET operating curves have an ohmic region, pinch-off region, and avalanche region. Operation in the pinch-off region will make drain current independent of source/drain voltage fluctuations. The family of curves is used to establish operating parameters. ☐ Bias for the FET can be fixed bias, source bias, gate-leak bias, or divider bias.

review questions

1. Does the common-source circuit have gain?
2. Which circuit configuration can have no gain?
3. Describe the input and output phase relationships of the three circuit configurations.
4. What is another name for the common-drain circuit?
5. Name the three regions of a FET operating curve.
6. Which region should the FET operate in?
7. What can happen if the FET is driven into the avalanche region?
8. How does source bias work?
9. Describe three forms of fixed bias.
10. What will happen to the output signal if a bypass capacitor is not used with some bias methods?

integrated circuits

By far, the greatest innovation in the field of semiconductor electronics is the advent of *integrated circuits* (*IC's*). From what you learned earlier in this book, you know that the semiconductor manufacturing process can be controlled by doping, which produces transistors and diodes with various resistance and junction characteristics. The doping process can also be applied to an ordinary piece of semiconductor material to give it a particular resistance; and the junction capacitance of a p-n diode can be manufactured for various values of capacitance. Thin films can also be deposited on a *semiconductor chip* to produce resistors or capacitors in much the same way that ordinary discrete resistors and capacitors are made. IC's, then, can be made so that resistors and capacitors are manufactured within or on the same semiconductor block of material. They can be manufactured simultaneously with the transistors and diodes.

This type of circuit manufacture eliminates the need for the large amount of wiring and mounting hardware used for conventional circuits. Also, since the individual components in an IC do not have to be handled in building the circuit, they can be produced microminiature in size. It is not unusual, for example, for an IC chip that is about 60 × 100 mils in size to contain about 350 components. This is a great improvement in size and weight—and also cost—over conventional hand-wired and *printed wiring* circuits. The optimum density of the number of individual components per IC chip is limited by manufacturing cost. With ordinary IC's, the optimum density is considered to be approaching 100. But, with *large-scale integration* (*LSI*), component densities of 1000 per chip are being sought.

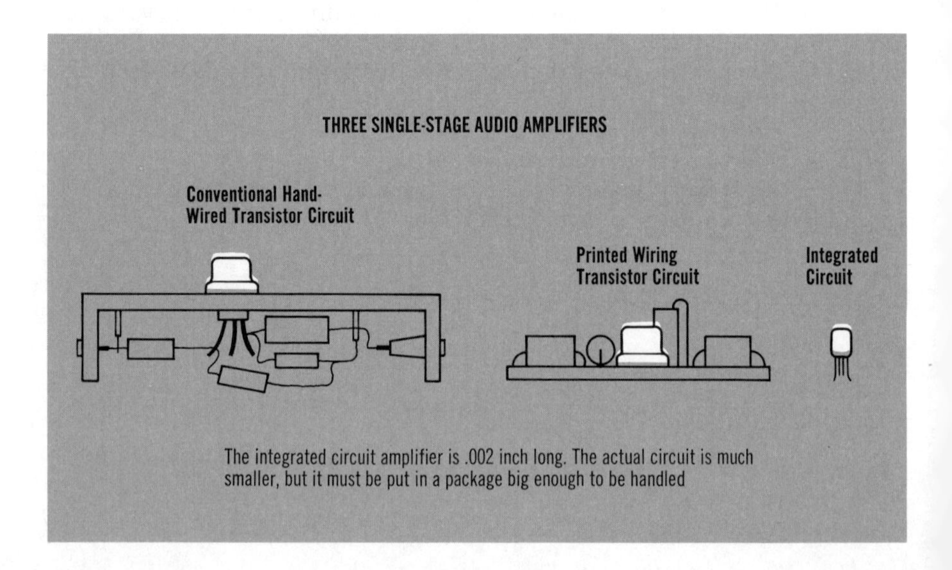

The integrated circuit amplifier is .002 inch long. The actual circuit is much smaller, but it must be put in a package big enough to be handled

monolithic and hybrid circuits

The ultimate aim in IC design is to incorporate any complete circuit, or group of circuits, or even a complete piece of equipment on one solid semiconductor slab. Whenever this is done, the product is called a *monolithic IC*. However, this is not always possible because it is not yet feasible to fabricate certain resistance or capacitance values, and in addition, it is not yet possible to manufacture many other parts by the semiconductor process. Naturally, coils and electrolytic capacitors are difficult to produce, as are all electromechanical parts. For many applications, then, an incomplete IC is combined with ordinary discrete components. Also, for manufacturing efficiency, it might be too costly to group all of the required circuits on one monolithic block; so separate chips are made, which are then connected together externally. Whenever IC's have to be connected to other components to accomplish a function, the overall circuit is called a *hybrid IC*.

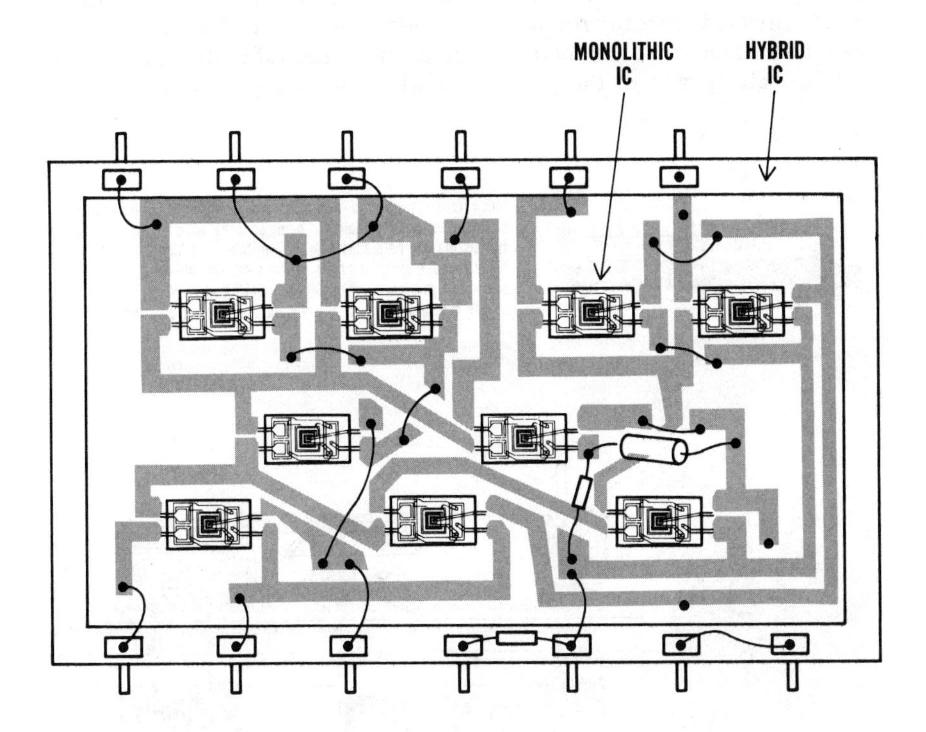

A monolithic IC can fulfill its purpose without the need for external parts. All of the circuit components and connections are contained in one monolithic block of semiconductor material

A hybrid IC combines monolithic IC's with external parts. This hybrid IC contains 9 monolithic IC's on a 1 × 0.8 inch board

mesa transistors

The idea for IC manufacture evolved when the art of manufacturing transistors reached the point where it became feasible to control semiconductor characteristics at reasonable costs. But it was the development of the mesa technique, which then led to the planar technique, for manufacturing transistors that first made IC's practical. These techniques allowed numerous components to be grown simultaneously on one semiconductor wafer; whereas, the earlier transistors were grown individually. *Mesa junction transistors* are made by forming one semiconductor segment on top of another. First, a number of isolated collector segments are grown simultaneously around the wafer; then base segments are all grown simultaneously *on* the collectors. Lastly, the emitter portions are all grown simultaneously in the bases. The result is a large number of transistors made simultaneously on the same wafer, which can then be sliced up to separate all the transistors. IC's can be (and have been) made this way, but since all of the elements end up as plateaus or raised "islands" on the wafer, they have to be interconnected with microminiature leads. This is a tedious and expensive task. The mesa method is still used to manufacture transistors, but has given way to the planar technique in the production of IC's.

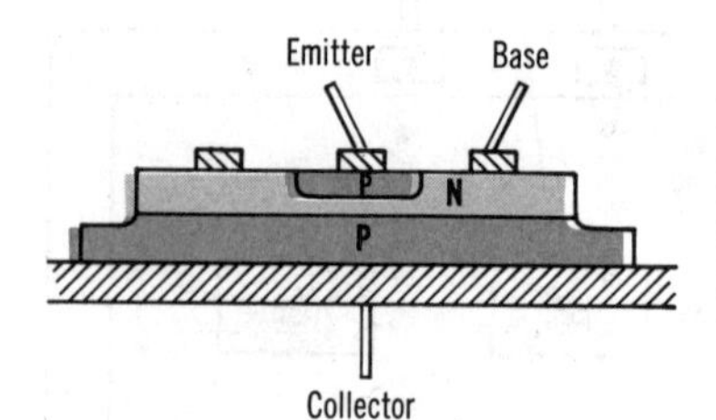

Mesa transistors are built up on a common wafer to form plateaus or raised islands. Typically, as many as a few hundred mesa transistors can be manufactured simultaneously on a 3/4 inch disk. Because sections grown by the mesa technique are separated, microminiature interconnecting wires must be used to join the sections if an IC were made by the mesa method

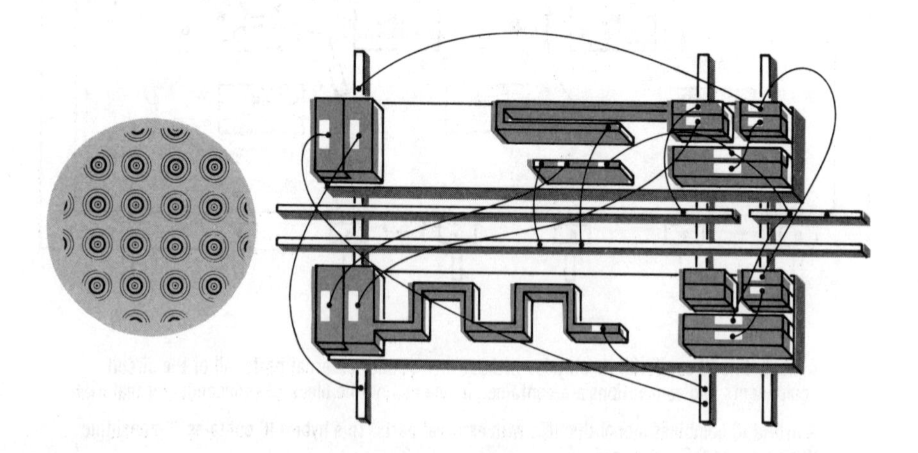

planar transistors

Planar transistors are manufactured in quantity on the same wafer just as the mesa transistors are, but they are all produced horizontally in the same plane on a continuous slab. The material is treated to form the doped transistor segments within or separated from one another, and the wafer ends up as a flat continuous slab that can be sliced up into individual transistors. For IC's, the planar technique is ideal, since other components can be produced in the wafer already joined to the part it is supposed to be connected to in the circuit. Also, since the planar wafer is always a flat continuous material, printed wiring and printed components can be easily applied to the surface to complete the circuit connections. A typical example of a planar IC is shown on this page. The geometry is arranged so that it is easy to understand. A more realistic geometry for this IC is shown on page 4-146.

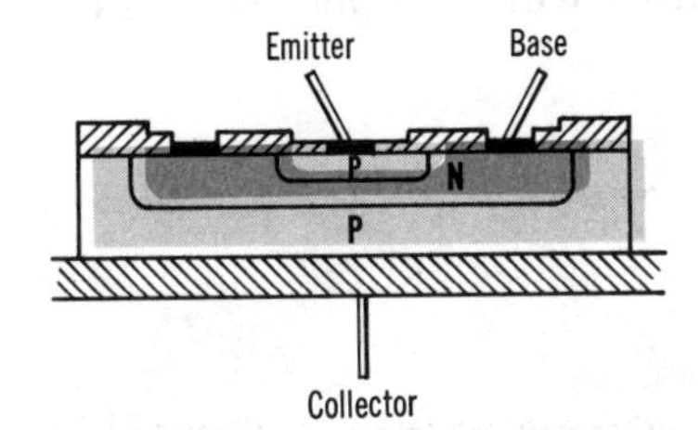

Since the planar technique depends on changing positions of an already built-up semiconductor block, the sections are all produced along the same lateral plane. The segments can be joined, either within the wafer, or with printed wiring. This is a simplified example of an a-f amplifier IC

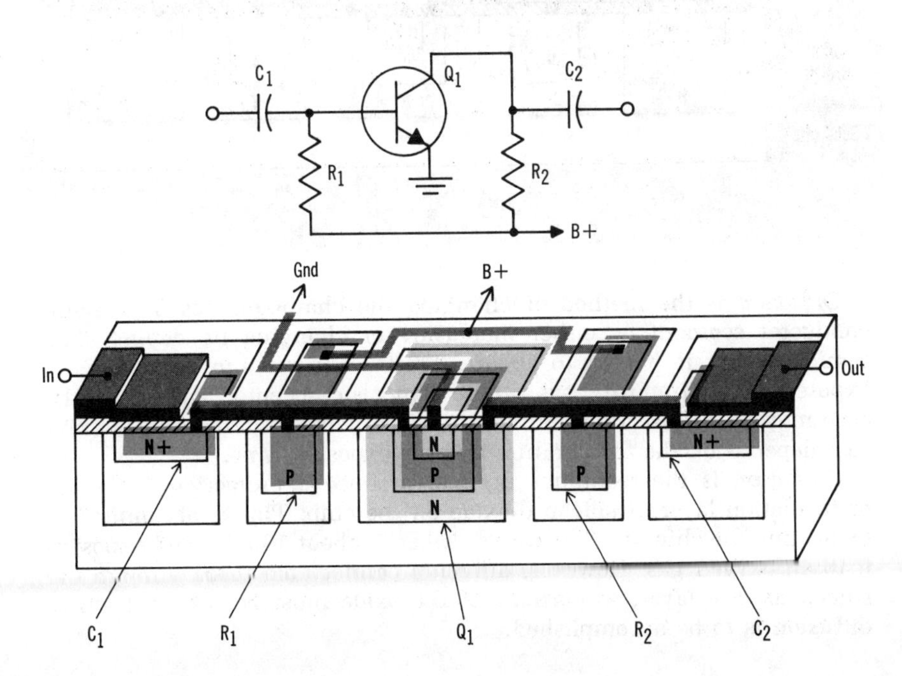

manufacturing processes

There are numerous techniques used during the various phases of IC manufacturing. The following are descriptions of the more typical reliable methods. Page 4-139 shows a step-by-step example of how a basic IC is produced.

All IC chips start with a pre-manufactured *substrate*, upon which the IC is built. The substrate provides support and a means of isolating various IC components. This is generally a doped silicon chip of the proper size to hold the IC components.

Epitaxial growth is the technique used to develop the semiconductor segment that will be given the various component characteristics. This is generally done by heating the substrate to about 1200°C in a vacuum chamber, and exposing its surface to a mixture of gases that solidify to form a junction and grow on the substrate. Generally, a growth of about 25 microns thick is produced. For p-type growth, a mixture of silicon gas and an impurity gas such as boron, gallium, or aluminum is used. For an n-type growth, silicon gas is used with phosphorous, antimony, or arsenic gas.

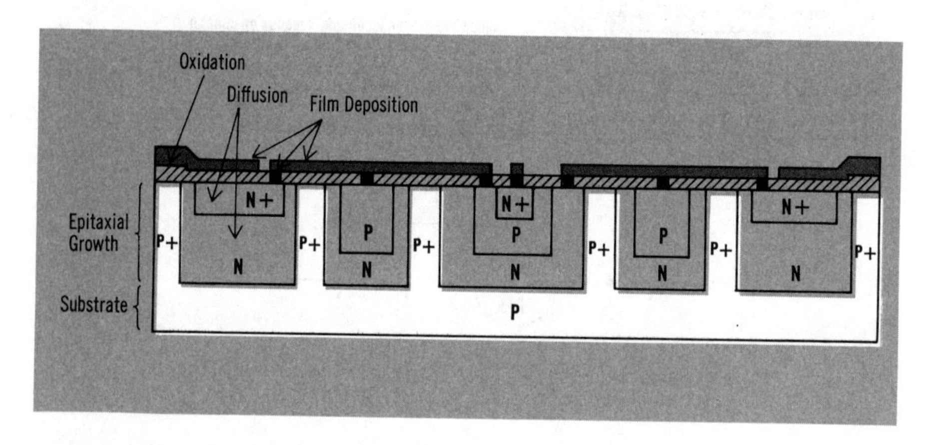

Diffusion is the method of changing the characteristics of a semiconductor segment by either increasing or changing its doping. The semiconductor is heated to about 1000°C in a vacuum chamber and exposed to an impurity gas whose donor or acceptor atoms, as the case may be, diffuse into the material. The depth and amount of diffusion depends on the temperature and the exposure time.

Oxidation is the technique used to produce a protection, isolation, or insulation layer of silicon dioxide on the chip. This is accomplished by heating the chip in a vacuum chamber to about 1000°C and exposing it to an oxygen gas. However, diffusion cannot take place through the silicon dioxide layer, so portions of the oxide must be removed where diffusion is to be accomplished.

manufacturing processes (cont.)

Photolithographic etching is used to determine which segments of the oxide layer are removed. This is accomplished by first coating the entire oxide layer with a *photosensitive emulsion*, such as Kodak KPR *photoresist.* The emulsion is then exposed to ultraviolet light through a stencil-type mask that blocks out light from the parts of the oxide that must be removed. The mask geometry is actually photographed onto the photoresist emulsion. The photograph is "developed" by applying a chemical, typically trichloroethylene, which dissolves those parts of the emulsion that were *not* exposed to light. At the proper time, a *fixing* solution is applied to stop the developing action. Thus, a pattern of photoresist film is left that covers and protects those portions of the oxide layer that should not be acted on by any other chemical process.

Chemical etching is used to dissolve the oxide layer from those areas where the photoresist film was removed. Etching is usually accomplished by immersing the chip in hydrofluoric acid. The oxide layer is removed only from those areas where it was not protected by the photoresist film. The remaining film is then dissolved in a sulfuric acid bath and all residue is removed by abrasive polishing. This leaves a pattern of silicon oxide that exposes the semiconductor where diffusion should take place, and protects or covers those areas where diffusion should not take place. This permits *isolated diffusion.*

For further processing, the remaining oxide mask can be etched away and the entire process can be repeated to form a new pattern.

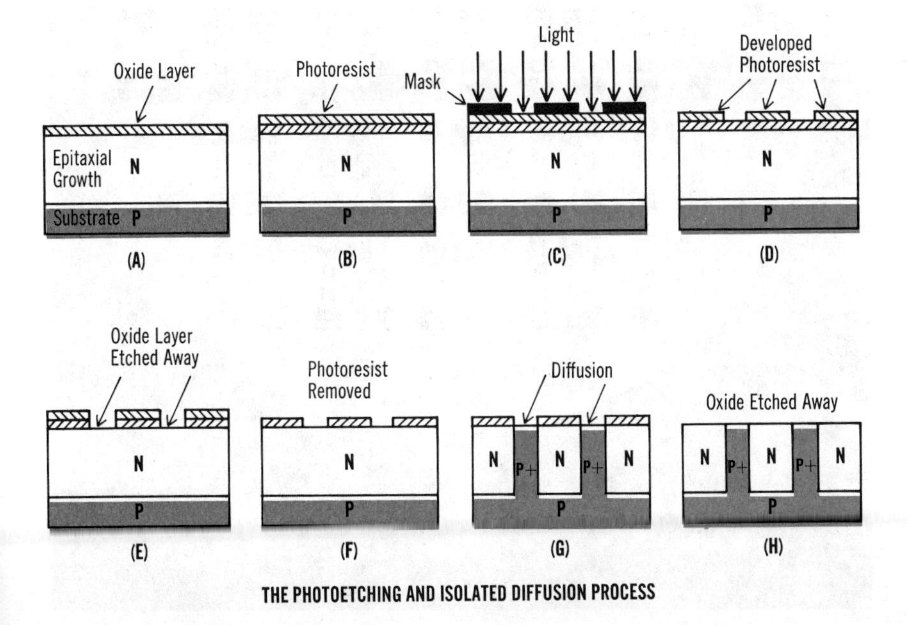

THE PHOTOETCHING AND ISOLATED DIFFUSION PROCESS

manufacturing processes (cont.)

Film deposition is the process of applying thin metallic films to act as terminals or wiring. Thin films are deposited on the chip in a heated vacuum chamber that allows an evaporated metallic film to condense on the surface of the chip. Then, the *photo etching* process is used to remove the parts of the film not needed.

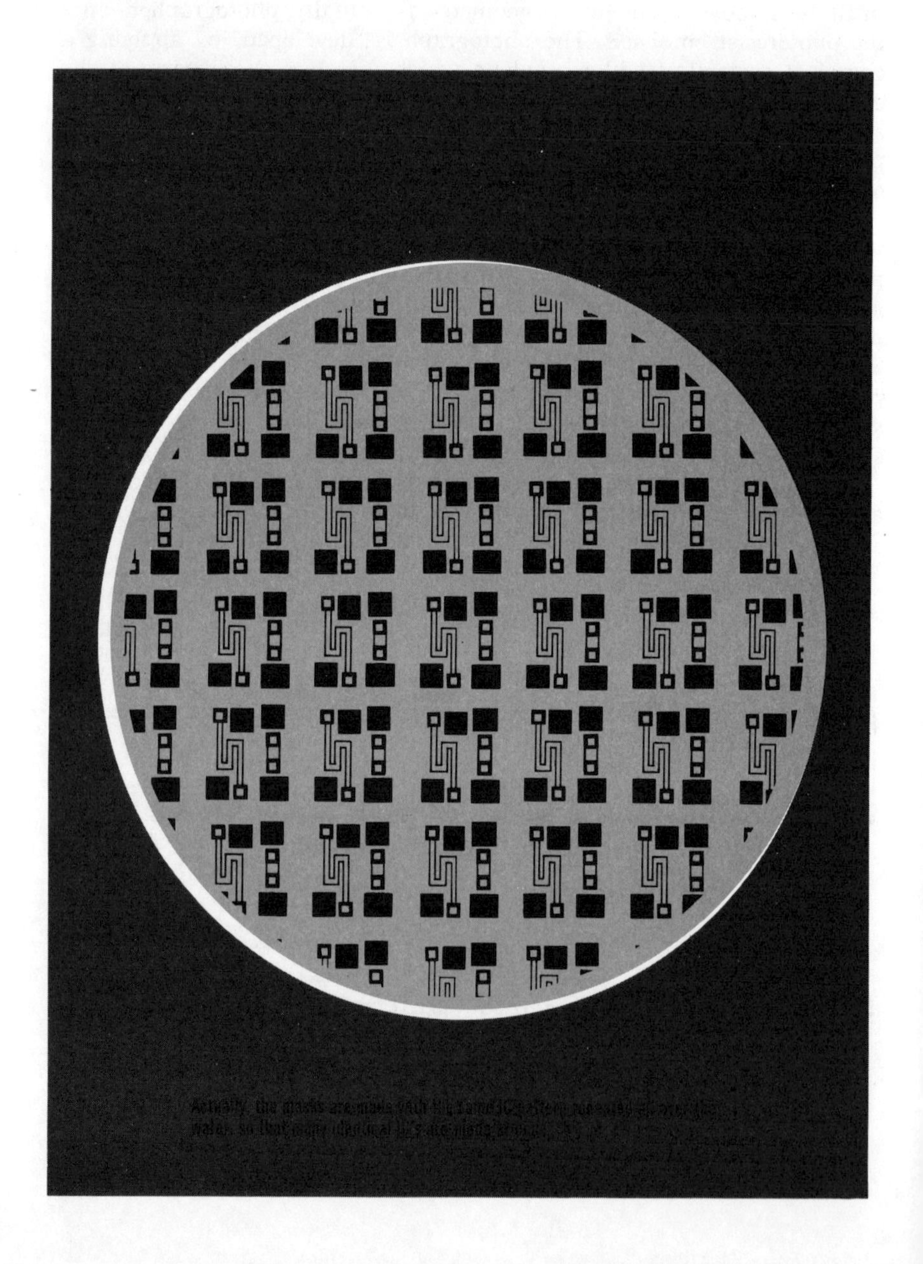

step-by-step sample

The following simplified example, shown on this page and the next, is typical of how an IC is manufactured. The circuit used in the example is the same as that shown on page 4-135, which gives a better overall view of the completed chip with all the terminals and wiring configuration. The geometry of the layout is simplified for teaching purposes. A more realistic geometry for this circuit is given later.

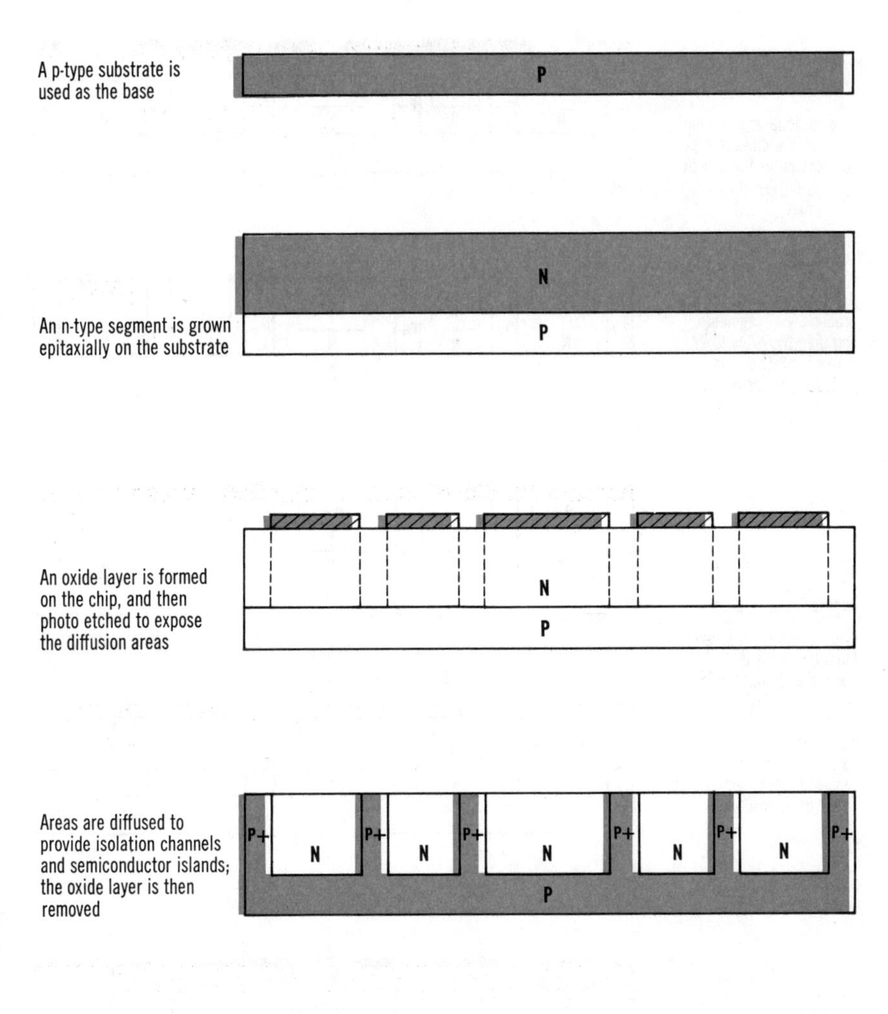

step-by-step sample (cont.)

Areas are diffused to produce p-segments within the n-segments. Then the oxide layer is removed

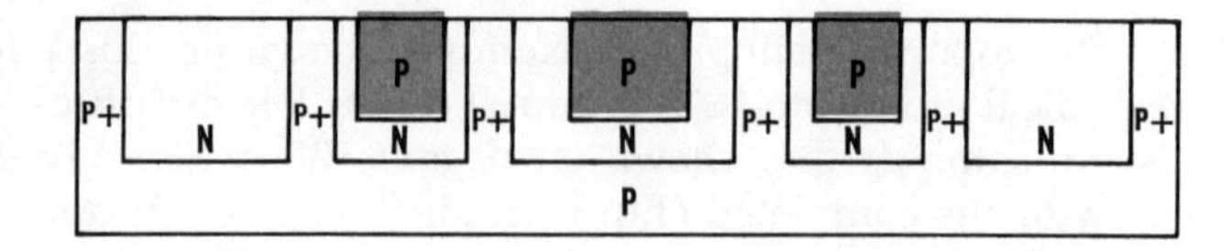

New oxide layer is formed and photo etched to expose the next areas for diffusion

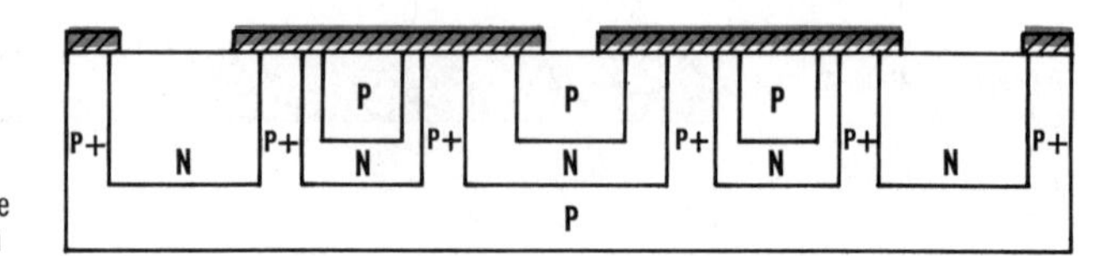

Areas are diffused to produce N+ segments, and then the oxide layer is removed. This completes the diffusion process

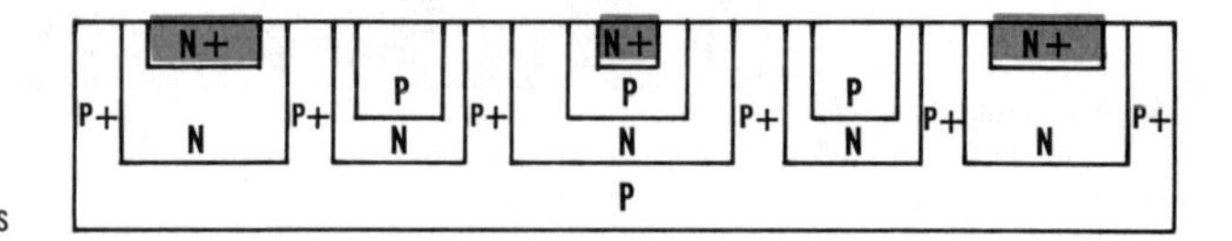

Oxide layer is formed and photo etched to expose terminal points

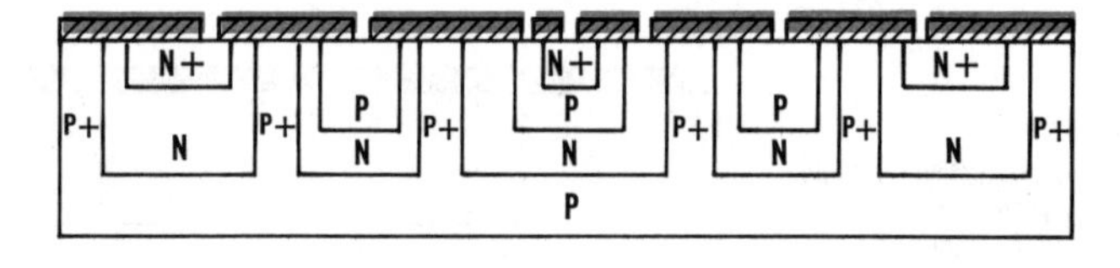

Metallic film is deposited and built up to form terminal connections on semiconductor segments. Oxide layer is not removed, but is left to act as an insulation layer

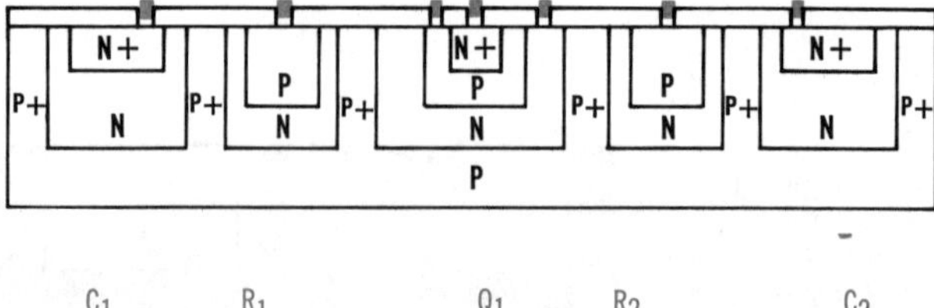

Metal film is deposited on the oxide layer to join all terminal connections. Then this film is etched away to form proper wiring patterns, capacitor plates, etc.

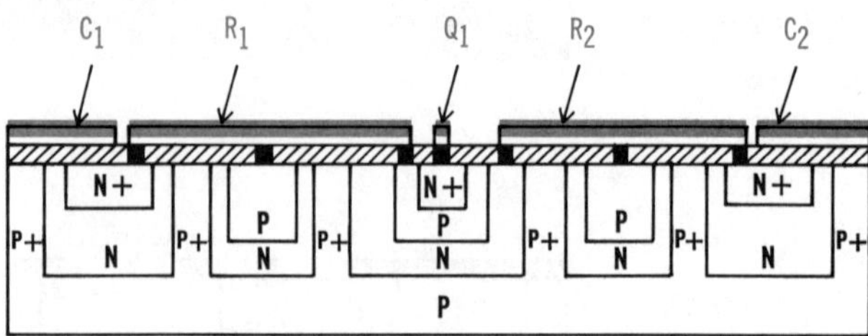

isolation

As you saw in the previous examples, the individual semiconductor "islands" were first produced in the epitaxial chip when heavily doped channels were diffused down to the substrate. These channels are given the opposite doping of the component island segments, and so form p-n junction diodes with them. When these diode junctions are reverse biased, they can isolate the segments from one another. This is important; since all of the components are part of a common semiconductor, there are natural paths for all sorts of unwanted couplings and feedback. To minimize feedback and coupling, the most negative potential of the IC is applied to the substrate to reverse bias the *isolation junctions*. These isolation diodes and their accompanying junction capacitances are a normal part of the equivalent circuit of any IC. They must be considered in any circuit design, since reverse biased diodes do conduct, and junction capacitances couple signals. As a result, *IC circuit design* is not the same as discrete component circuit design. An ordinary wired circuit cannot be directly converted to an IC design. IC design is based on an entirely different design philosophy. More details on *equivalent IC circuits* are discussed later.

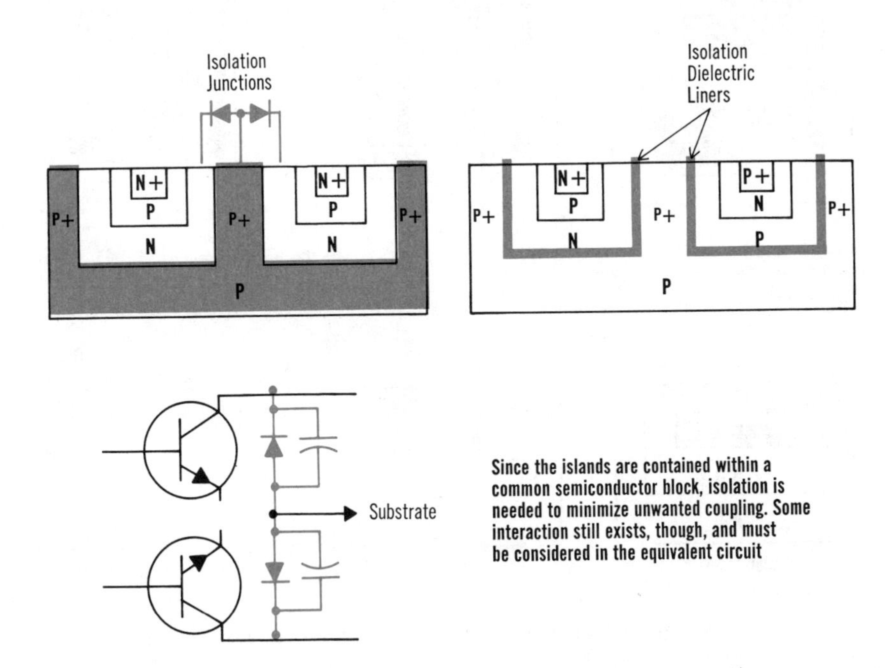

Since the islands are contained within a common semiconductor block, isolation is needed to minimize unwanted coupling. Some interaction still exists, though, and must be considered in the equivalent circuit

Another technique used to isolate IC components from one another is to line the islands with a dielectric or insulating material. This eliminates junction coupling, but capacitive coupling still exists. Some *dielectric liners* that are used are: silicon, oxides, ruby, and ceramics.

IC transistors

Whenever junction isolation channels are used, all the transistors on any one chip must be either n-p-n or p-n-p. They cannot be intermixed, otherwise it will be difficult to reverse bias all isolation junctions. The use of isolation liners, though, does permit a mix of n-p-n and p-n-p types.

The transistors shown in the previous diagrams were the bipolar junction type that function like the ones you studied earlier in this book. They are produced by diffusing each element within another, to form two concentric junctions. Although one terminal point can be used on each transistor element, each terminal is deposited around most of its element to reduce terminal contact resistance. In many cases, to further reduce contact resistance, a more heavily doped conductive channel is diffused in an element beneath the contact points. These are shown with a + sign, and are merely low-resistance semiconductor contacts.

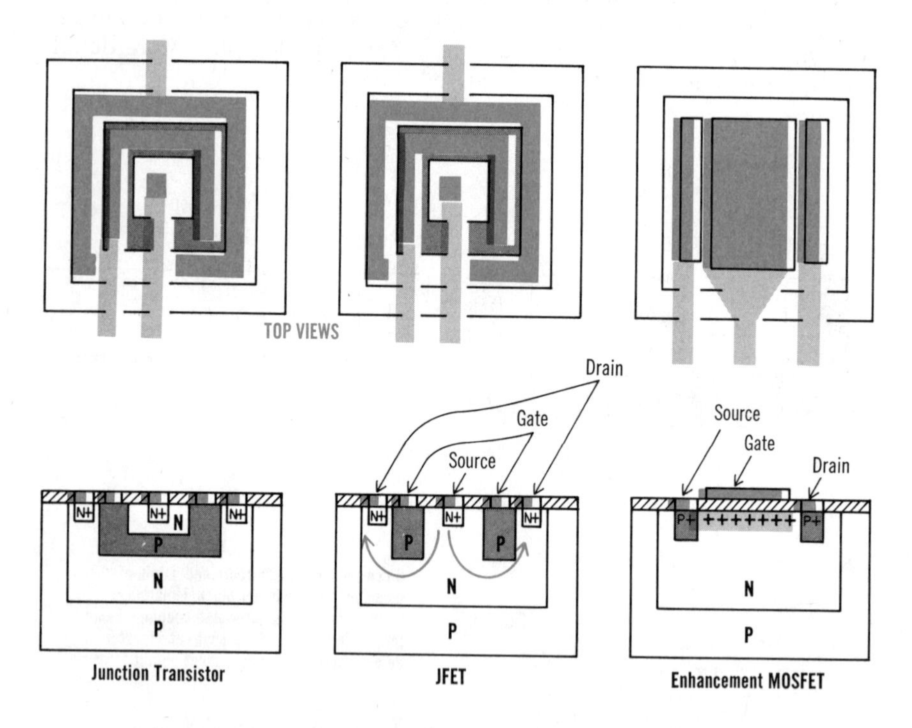

The N+, which indicates extra heavy doping termination segments, reduces terminal contact resistance. This is not usually important in the base

The diagrams on this page also show how the ordinary *junction FET (JFET)*, and *metal-oxide semiconductor FET*'s (*MOSFET*) are formed in an IC.

IC diodes

Many single diodes can be made in a manner similar to transistors, but only two elements are needed. Dual common-anode or common-cathode diodes are also easily formed, as shown on this page. As a matter of fact, the number of common-element diodes that are made is restricted only by the geometric pattern that is required.

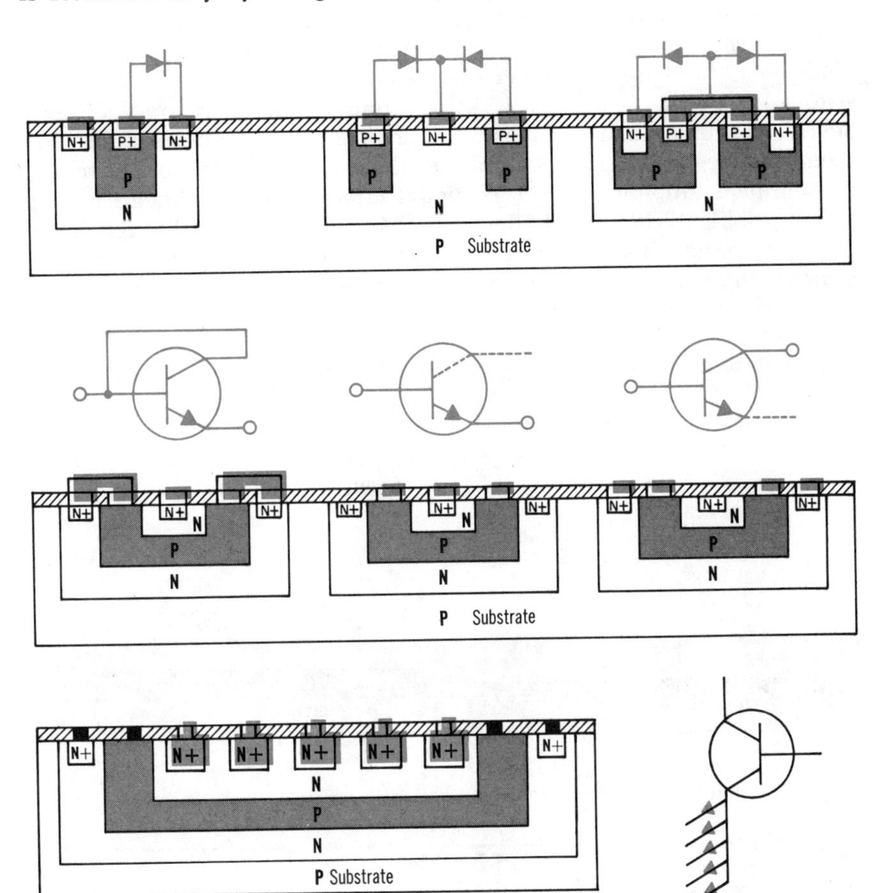

It is cheaper to repeat the same part on an IC than to make an IC with mixed components. So, it is not unheard of to use chips with nothing but transistors, and interconnect the chips to perform other functions. Since any two adjacent elements in a transistor form a p-n junction, those two elements can be connected as a diode.

To show the versatility and uniqueness of IC design, the diagram on this page illustrates a transistor with 5 emitters, which acts as a transistor with 5 diode inputs.

IC resistors

Resistance can be incorporated into an IC in four ways: (1) the *monolithic resistor* uses an area that is diffused into the chip with the proper dimensions and doping to produce the desired resistance, (2) a *metal film resistor* is deposited on the chip with the proper dimensions to produce the desired resistance, (3) a diode can be forward or reverse biased at the proper point on its operating curve to give the desired resistance, and (4) a transistor can be so biased, or used as a diode so as to cause the proper resistance. The use of diodes and transistors as resistors is difficult in some circuits, since they are nonlinear devices. Their resistance values could change when the current through them changes, since this depends on their operating curve. Their use in nonlinear amplifiers, such as digital circuits, is not critical; but in linear amplifiers they could affect response and fidelity of the signal. On the other hand, this nonlinear characteristic could be used to produce a *monolithic variable resistance*, or potentiometer effect.

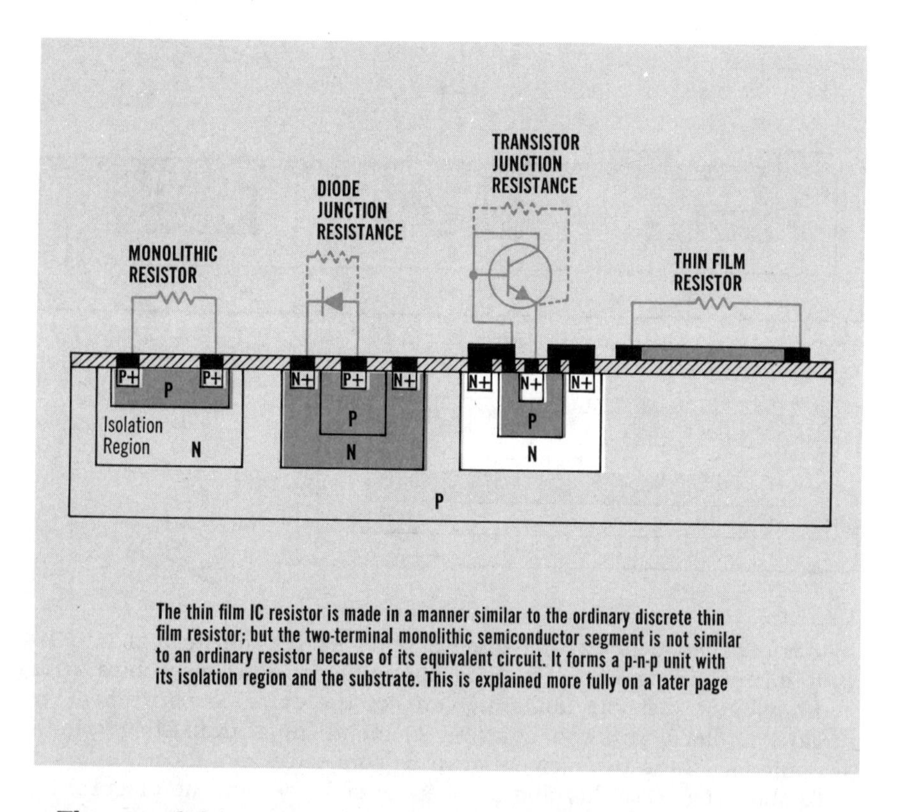

The thin film IC resistor is made in a manner similar to the ordinary discrete thin film resistor; but the two-terminal monolithic semiconductor segment is not similar to an ordinary resistor because of its equivalent circuit. It forms a p-n-p unit with its isolation region and the substrate. This is explained more fully on a later page

The monolithic resistor can also be doped to give it *thermistor* or *varistor* characteristics.

IC capacitors

Capacitance can be incorporated into an IC in five ways: (1) as a *metal-oxide semiconductor* (MOS) *capacitor*, (2) as a *thin-film capacitor*, (3) as a specially constructed *junction monolithic capacitor*, (4) as a p-n diode properly biased to produce the necessary junction capacitance, or (5) as a transistor used as a diode capacitor.

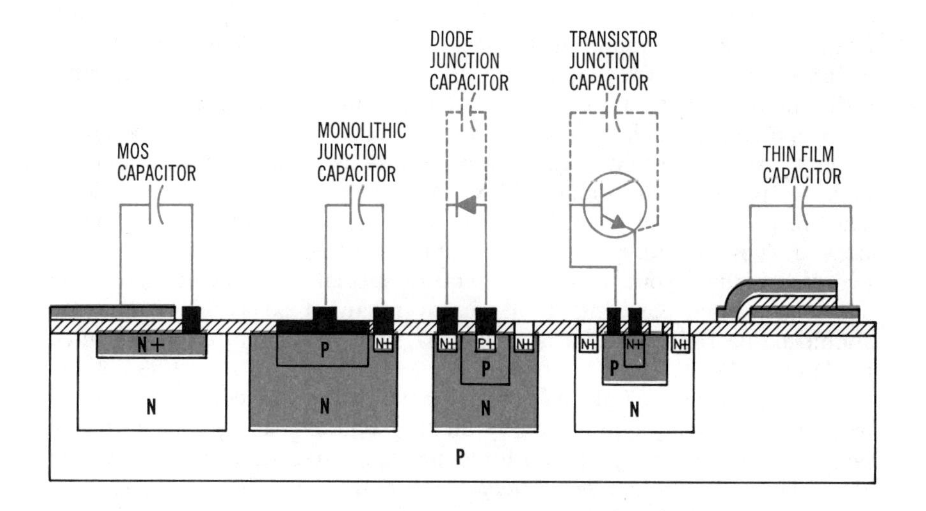

The thin film capacitor is made the same as an ordinary discrete thin film capacitor, with two thin film electrodes separated by a dielectric. The metal-oxide semiconductor (MOS) capacitor is similar, except that it uses a semiconductor segment as one electrode. The monolithic junction capacitor is made like a p-n diode, but with special geometrics and doping for controlled capacitance values

The thin-film capacitors are made in the same way as ordinary discrete thin-film capacitors, with an oxide dielectric separating two electrodes. The MOS capacitor is similar, but one of the electrodes is a semiconductor segment (N + in the drawing), and has a typical IC equivalent circuit. The other three techniques each use the capacitance produced across a p-n junction, but the diode and transistor were manufactured as a diode and transistor; they are biased for the needed capacitance. The junction monolithic capacitor, though it is constructed like a diode, was specially diffused and arranged for a junction that will create the proper capacitance. As with the resistors, only the thin-film capacitor is reliably linear. The other types could use a bias control that would offer some degree of variable capacitance. Also, an IC version of the *varactor diode* could be created, which is a p-n diode specially designed for variable capacitance.

layout, interconnections, and packaging

IC's are laid out geometrically for ease of manufacture, to reduce wiring interaction, and to obtain maximum isolation. Generally, the layout is arranged to balance the terminal connections around the chip, and to group the same kinds of parts together in common *isolation islands*. This type of grouping will reduce the number of isolation islands that will be required. However, each transistor requires separate isolation. When parts that occupy the same island are at considerably different potentials, the parts are grouped apart in their island according to potential. Each island is connected to the most positive potential of any part it contains, and the substrate is connected to the most negative potential to produce the reverse biased isolation p-n junctions. The thin-film wiring should be kept as short as possible and made as wide as possible to reduce its resistance. Crossover wires should be avoided in the layout since they require special oxide layer steps, and cause unwanted coupling. If a crossover cannot be avoided, the wire should be passed over a resistance area, since that is already protected by an oxide layer.

After the chip is finished, the external wires are connected to the terminals. This is done either by tacking pigtail leads to the terminal points or by creating thick film deposited wires that extend off the chip. These are called *beam leads*, since they can be strong enough to support the chip. The chip can be used directly on a *hybrid card*, or it can be packaged in: (1) a TO type of can, (2) a sealed flat pack, (3) a sealed in-line pack.

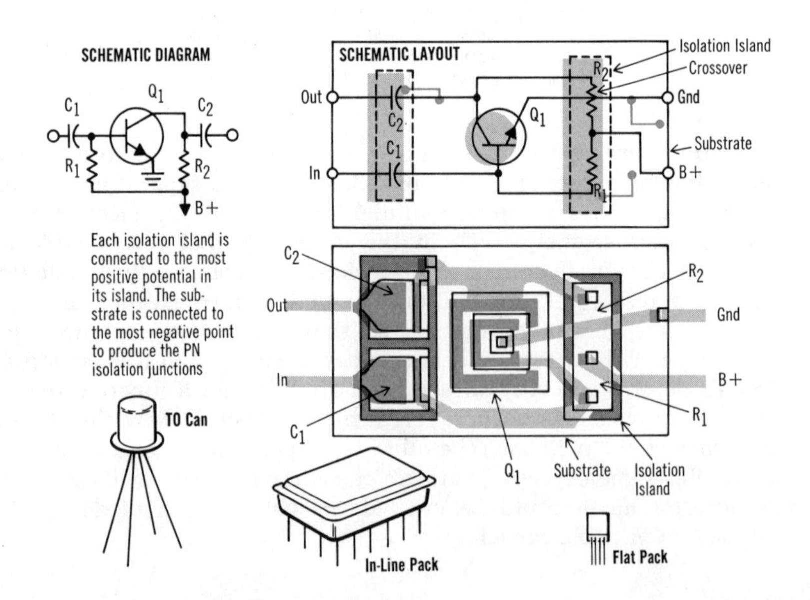

equivalent circuits

You learned earlier that various p-n junctions, capacitances, and resistances are added to any basic circuit because of the substrate, the isolation channels, and isolation islands: this is why IC design is approached differently. The diagrams on this page show the equivalent circuits of various IC components, and how the basic amplifier circuit that you have been studying for IC's is modified by the chip structure.

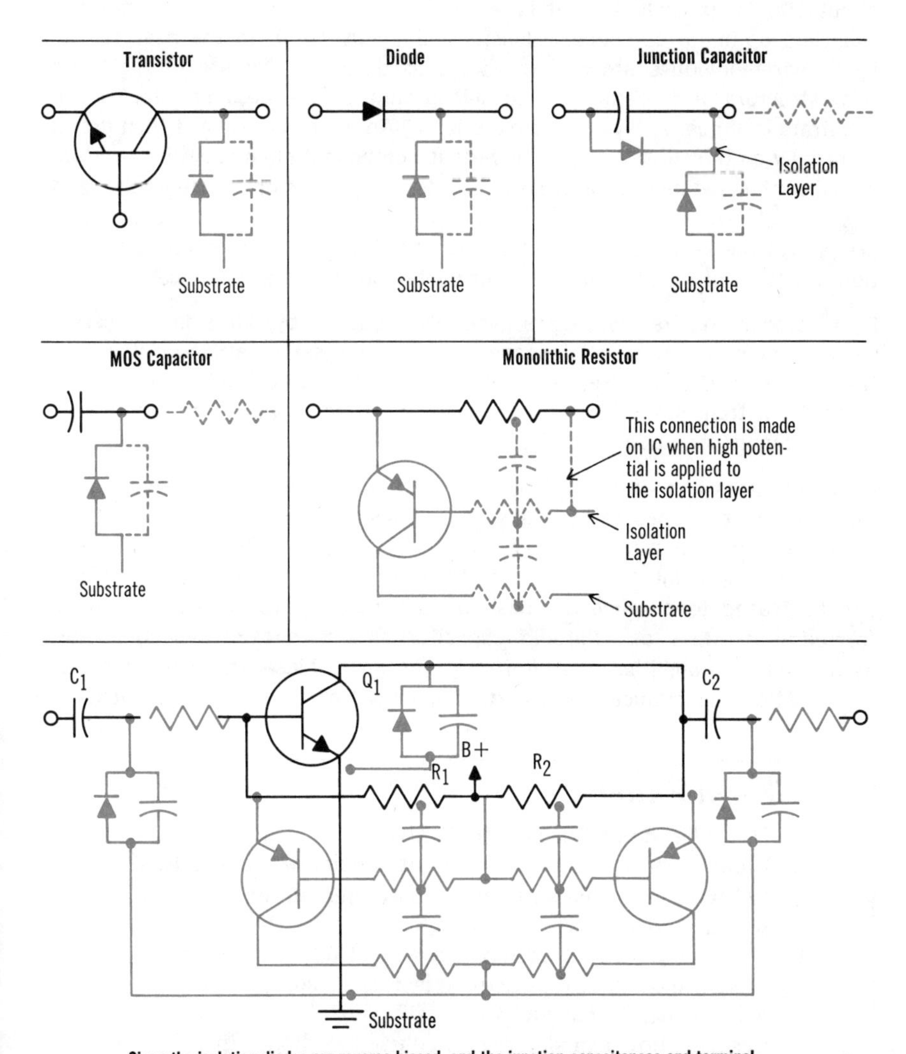

Since the isolation diodes are reverse biased, and the junction capacitances and terminal resistances are low, the basic circuit can function without much regard for the equivalent circuits, in many applications. But, with high-frequency, sensitive circuits, the equivalent circuits modify the basic circuits enough to warrant consideration

summary

☐ There are two basic types of IC's, monolithic and hybrid. ☐ In a monolithic IC, a complete circuit, or a group of circuits, is contained on one solid semiconductor slab. ☐ Because it is not yet possible to make many electronic parts by the semiconductor process, an incomplete IC must sometimes be combined with regular discrete components to accomplish a particular circuit function; when this is done, the overall circuit is called a hybrid IC. ☐ Considering manufacturing costs, the optimum component density per IC chip is about 100 for most IC's. ☐ It is expected that LSI (large-scale integration) will yield optimum component densities of about 1000 in the near future. ☐ Planar techniques are almost always used to manufacture IC's. ☐ The base structure of an IC is a doped silicon chip, called the substrate. ☐ The substrate is almost always p-type material. ☐ An n-type segment is then grown epitaxially on the substrate. ☐ The various component characteristics are produced in this segment by processes of diffusion, oxidation, photographic and chemical etching, and thin-film deposition. ☐ Because all of the components are part of the same semiconductor chip, the IC must be designed with isolation junctions or liners to minimize unwanted couplings and feedback.

☐ IC diodes and transistors are generally produced by diffusion processes. ☐ In many cases, since it is cheaper to repeat the same part in an IC rather than make an IC with mixed parts, it is quite common to produce an IC with nothing but transistors. The transistors can then be connected to function as other components, such as diodes.

☐ IC resistors can be formed by four methods: a block of semiconductor material can be doped to have a specific resistance; a thin film can be deposited on an IC chip to produce a specific resistance; a diode can be biased at an operating point that will provide the desired resistance; and a transistor can be biased to produce the desired resistance. ☐ IC capacitors can be formed in five ways: as a thin-film capacitor; as a metal-oxide semiconductor; as a junction monolithic capacitor; as a p-n diode biased to produce a certain junction capacitance; and as a transistor operated as a diode capacitor.

review questions

1. Describe the two basic types of IC's.
2. What is the optimum component density of modern IC's?
3. What is the basic manufacturing technique used in the production of IC's?
4. Describe the construction of a typical IC.
5. Describe the methods used to isolate IC components.
6. What is the "substrate" in an IC?
7. Describe how transistors can be used as diodes in IC's.
8. Describe four ways of obtaining resistors in IC's.
9. Describe five ways of obtaining capacitors in IC's.
10. What does the term "LSI" mean? What component densities can be achieved through LSI?

index